ROUTLEDGE LIBR
GEOLO

Volume 10

THE GEOGRAPHY OF SOIL

THE GEOGRAPHY OF SOIL

BRIAN T. BUNTING

LONDON AND NEW YORK

First published in 1965 by Hutchinson & Co (Publishers) Ltd
Second edition 1967

This edition first published in 2020
by Routledge
2 Park Square, Milton Park, Abingdon, Oxon OX14 4RN

and by Routledge
52 Vanderbilt Avenue, New York, NY 10017

Routledge is an imprint of the Taylor & Francis Group, an informa business

British Library Cataloguing in Publication Data
A catalogue record for this book is available from the British Library

ISBN: 978-0-367-18559-6 (Set)
ISBN: 978-0-429-19681-2 (Set) (ebk)
ISBN: 978-0-367-20355-9 (Volume 10) (hbk)
ISBN: 978-0-367-20710-6 (Volume 10) (pbk)
ISBN: 978-0-429-26301-9 (Volume 10) (ebk)

Publisher's Note
The publisher has gone to great lengths to ensure the quality of this reprint but points out that some imperfections in the original copies may be apparent.

Disclaimer
The publisher has made every effort to trace copyright holders and would welcome correspondence from those they have been unable to trace.

THE GEOGRAPHY
OF SOIL

Brian T. Bunting

Lecturer in Geography, Birkbeck College,
University of London

HUTCHINSON UNIVERSITY LIBRARY
LONDON

HUTCHINSON & CO (*Publishers*) LTD
178–202 Great Portland Street, London W1

London Melbourne Sydney
Auckland Bombay Toronto
Johannesburg New York

First published 1965
Second edition 1967

*Cover design of paperback edition by
courtesy of W. G. Gale/Barnaby*

*This book has been set in Times New Roman,
printed in Great Britain on Smooth Wove paper
by Anchor Press, and bound by Wm. Brendon,
both of Tiptree, Essex*

Contents

Figures

Tables

Abbreviations

Units: The metric system is used.

Chemical symbols are also used as abbreviations.

Horizons are usually referred to as A2 or Bh, not A_2 or B_h.

The original spellings are usually used in quotations or on diagrams. Hence the dualism of 'podsol' and 'podzol', and variations on 'ferrallitic' and on 'grey'.

Abbreviations in the text or on diagrams.

ø	diameter (particles)
Ø	mean annual (eg rainfall)
w	weight
v	volume
ppm	parts per million
m p gm	millions per gram
μ	0·001 mm
Ft	ferrallitic soil
Fc	ferritic soil (= ferrisol)
Fg	ferruginous (tropical) soils (= fersiallitic soils)
BP	before present (ie before 1950)
CEC	cation exchange capacity
>	greater than . .
<	less than . .
g/b	grey-brown
r/y	reddish yellow
bl	black
grn	green
m	mottled
dk	dark
lt	light

fn	fine	lm	loam
cl	clay	sd	sand

Preface

The position of soil studies in geography, especially in Britain, is strangely nebulous and unsatisfactory. Many admirable texts on soil science are available, intended for students of agriculture, yet the essence of soil geography—the study of the distribution and morphology of soil in relation to external influences and internal processes—is less well represented in them.

I have tried to present the results of modern national and international soil studies within—but only just within—the traditional framework of the 'Great Soil Groups', and to take stock of the soils of the world on a regional, rather than on a zonal, basis. I have also preferred to use detailed examples and quantitative data, though a whole Institute is needed to carry out such work with both adequacy and justice.

While I have been writing this book new horizons have been opened in Soil Science. The Seventh Approximation and the translation of *Pochvovedenie*, as *Soviet Soil Science*, have appeared. So, too, have the Israeli translations of several major Russian works, as well as Professor Mückenhausen's elegant studies of the soils of Germany. Apart from modern works from many lands I have also turned to some of the pioneer works on soil, both as sources and for inspiration.

Acknowledgment of sources is difficult in a work such as this, which necessarily greatly depends on the work of others. I trust that errors of omission will be excused and that errors of simplification or of understanding will be corrected as gently as possible. The removal of a few inconsistencies in soil nomenclature seemed a desirable though risky undertaking, and to avoid seeming contradiction I have occasionally bent the hypothesis of X to fit the later observations of Y.

I wish to thank many colleagues for help and advice, and to acknowledge the kindness of more distant workers in supplying studies and reports. I am indebted to Mr B. W. Avery, Mr R. D. Green and Mr B. Clayden for their comments on parts of the text, to Dr A. Young for comments on Chapter 17 and to Mr H. C. Moss, of Saskatchewan, for discussion on Canadian soils.

I also wish to remember the example set by Professor R. S. Waters, who, when first at Sheffield University from 1952–54, gave courses in soil geography from which a number of workers on soil have stemmed; also to thank Hr N. K. Jensen and Jens Kragelund-Øvig of Viborg, for providing a rigorous training at an impressionable age.

To these, and to my young son Mark, whose enthusiasm for digging I endeavour to match, I dedicate this book.

BRIAN T. BUNTING

Birkbeck College

THE
GEOGRAPHY OF SOIL

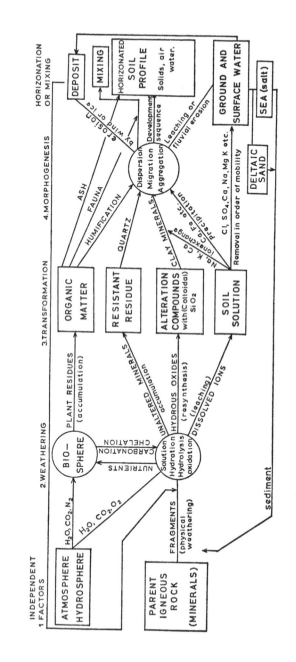

Fig. 1 A scheme of soil profile development (after Yaalon, 1960)[1]

1

Introduction

W HEN a rock is first revealed at the surface it is subjected to the action of atmospheric and biologic agents. The first stages in its alteration to soil are achieved through increasing fragmentation by physical weathering, and by hydration and other more complex forms of chemical weathering. These of themselves do not create a soil, only a 'weathering mantle'—the parent material of soil. Simultaneous with this breakdown of the rock, some of its easily soluble components are involved in the nourishment of invading micro-organisms and these, as time passes, increase in bulk, complexity of life form and in their effect on the mantle.

The uppermost parts of the mantle are most affected by such organisms, and by the succeeding higher plants, and these parts are also most susceptible to the daily and seasonal variations of climatic factors. It is the climatic factors which first influence the rate of formation of the weathering complex, then the living conditions of invading plants and organisms, and finally both together—climate and plants—control the formation and differentiation of soil horizons to produce a soil profile[1] (Fig. 1).

The first steps towards an understanding of soil are therefore (1) a study of the possible origins and components of rocks and deposits from which the soil has been derived; (2) an assessment of the factors of climate significantly active in soil formation; (3) an examination of the forms of organic material capable of interacting with the soil or its parent material. By implication one also needs to assess the period of time since a 'time zero', when the new surface was first exposed.[2] Examples of this 'time zero' need not involve a direct geologic or tectonic cause, for many are concerned with geomorphology; a surface newly freed from beneath an ice cover,[3] a newly emerged coast plain, or a new deposit of loess, provide surfaces on which soils will form.

The first stages of biotic action on a rock surface are evident in the most unpromising conditions of intense cold.[4] Finds of algae, bacteria and fungi are reported in the near-glacial environments of Antarctica[5] and north Greenland and in the nivation zone of high mountains. Lichens may colonize rock surfaces already prepared by algae and fungi and disrupt the rocks both by micro-physical and chemical action.[6] This gives increased access for moisture into the rock and a mass of dead organic material forms for later colonization by higher plants. Intense cracking of rocks by physical agents may form unstable areas which microflora cannot colonize, or it may form sheltered micro-environments in which fine mineral matter collects and in and on which mosses, grasses and annual plants can thrive.

The first stages of rock decay result in the removal of easily soluble elements from rocks (Cl, S, Ca and Na) which enter into the plant or are lost by leaching—the downward percolation of soil solutions. At the same time a relative increase in the less soluble minerals (Mg, Si, Fe and Al) occurs in the fine material of the upper part of the weathered mantle and quartz, and organically derived C, N and P also increase.[7] By the time grasses and shrubs are established, for example in a subarctic environment, the appreciable depths of fine earth formed show marked horizonation: the uppermost part darkened by organic matter (Ah), overlain by a mat of plant residues (L, F, H) and underlain by resistant, light coloured, inorganic material depleted of soluble minerals (A2). The whole may be termed an A horizon in which only a few fragments of the original parent material may survive, surrounded by weathering skins.[8] These Dokuchayev termed 'd' material. Often the A horizon has been much affected by geomorphic agents (Au) and may not have developed *in situ* from the parent material. It may also have been disturbed by cultivation (Ap) or have wind transported material added to it (Aa).

Deeper still, a B horizon has less organic material, more and larger fragments of parent material, and frequently shows redeposited material derived from the A horizon. This material may have been transported downwards unaltered ('in bulk') by simple physical processes or by biologic agents. Yet it can be material resynthesized by chemical means from the soil solution to form secondary minerals.[9] A lower layer, the substratum, is more compact; the decomposed or shattered regolith, known as the C horizon, or (more with hope than conviction) the parent material.[10]

The A horizon is considered as one of 'eluviation', the B as one of 'illuviation'. However, these process complexes are not mutually exclusive. The A horizon receives mineral and organic matter from plant decay, rainfall, wind blown dust, animal burrowing, and from manures, mulches and fertilizers on cultivation. The B horizon, though dominantly receiving minerals washed out of the A horizon, still loses the most readily soluble materials by leaching to the

Fig. 2 Dokuchayev's illustration of horizonation

drainage waters under moist climatic conditions on well drained sites. Difficulties also arise with the C horizon, for its climate is very different from the atmospheric and surface climates influencing the plant cover and A horizons. As we shall see, most soils have been characterized by the appearance of their upper horizons, not of their parent material. Generally, then, one may consider the soil as *the result of the modification of some loose mineral mantle by geographic agents such that distinct horizonation of the material occurs.*

The concept of horizonation is central to all field studies of the soil and each soil may be categorized by the degree of development of its vertical cross-section. This, called the *soil profile*, becomes more clearly developed with the passage of time (Fig. 2). Descriptions of the observable properties of each horizon—for example, colour, depth and texture—coupled with analyses of inferred properties such as permeability, pH and base saturation, form the basis of many recent scientific soil classification schemes. For geographic purposes, a scheme which relates these properties and the total profile to the environmental factors responsible is still to be preferred, for the main aim of geographical studies of soil is not to take the soil profile from its environment and place it in a completely integrated scheme of classification, but to consider it as the synthesis of past and 'present' environmental influences acting on a small and discrete part of the earth's surface. Yet classification is a valuable exercise if only as an aide-mémoire.

The soils of the world have been considered as belonging to one of the three major *orders*—zonal, azonal and intrazonal—each divisible into many groups.[11] Where the profile reflects the influence of the active factors of climate and organisms, over considerable periods of time, on the essentially passive factors of parent material and relief, the soils developed are called *zonal* soils. Each is *ipso facto* representative of a particular zonal climate and possesses an appropriate horizonation and clay mineral assemblage.

If the original parent material has been exposed to the active soil forming factors of climate and organisms for only a short time, or if it is derived from rock which is very resistant to weathering, or if it is sited on an unstable slope, then the soil has but poorly developed horizons and is known as an *azonal* soil. The influence of parent material and minimal time are always evident and such soils have very few related properties from group to group—even though their sites and situations are globally very similar.

In depressions where ground or seepage water is at or near the surface for long periods, or on plateaux where internal vertical and external lateral drainage are restricted and erosion absent, soils develop distinctive profiles not possessed by the nearby zonal soils and are known as *intrazonal* soils. Though reflecting the dominance of some local drainage factor, it is apparent that they are often geographically associated with zonal soils. In the recent American classification—the Seventh Approximation—they are considered as *aquic suborders* of each order of soils. In addition there are distinctive intrazonal soils developed on limestone or on basic volcanic rocks, in which parent material dominates the dynamic factors for considerable but not unlimited time. There are, too, anthropomorphic soils of the intrazonal order.

Most of the factors influencing soil formation have been studied by geographers, but not directly in relation to the soil profile. Probably this is because it is only a narrow representative of a more extensive body or area of soil—the soil 'site' or landscape unit. This, though essentially similar in vertical section to the profile, may be thought of as a large sheet or lense of soil, irregular in shape, either changing gradually to another profile, or else ending abruptly at some change of slope.[12] In characterizing these sites or slope elements, therefore, geomorphology may link with soil science, for age of site and intensity of weathering are germane to both sciences.

Many soils have been affected by man through cultivation or irrigation accelerating the leaching of minerals, cultivation mixing the uppermost layers and so concealing local differences in the surface horizons. Soils often come to resemble one another more closely after cultivation, and so man may be considered as a factor in soil formation. Similarly one may not assume that soils are always or solely in equilibrium with the current factors of formation. Some soils are related more to past events and are not interacting with the presently operating factors. Other soils, though developing, are lagging behind the environmental influences, an effect likened to hysteresis by Milne. Yet others are constantly interacting with the environmental factors which are not always in equilibrium; hence their effect, therefore, often changes in intensity, by mutual interaction, within short periods of time.[13]

As soils come to be surveyed more extensively, especially in tropical areas where soils have developed without interruption over greater periods of time, it is becoming clear that a monocyclic concept of soil formation is not the only possible interpretation.[14] In this monocyclic view the soil comes to be in a state of balance with the external factors, especially of climate, to form a climax soil. We now know that many soils are polycyclic in nature and origin. Their orderly development has been interrupted by changes of climate (hence of weathering intensity) or by the more abrupt intervention of geomorphic agents giving truncation of the profile on erosion or else burial by deposition of fresh material upon the pre-existing soil. Thus a single soil profile may bear witness to one or even all three of these happenings and records a succession of environments, and the geomorphic history of the landform on which it is sited.

Although the soils of the world appear to be distributed in the form of zones corresponding to climate, or as azonal or intrazonal soils responding to local circumstances, the impression of zones can only be validated by many successive stages of generalization in compiling a small scale distribution map for the soils of the world or of a continent. In reality the detailed variations of profile within a zone are far more instructive than these generalized zones and groups and are more clearly related to factors other than zonal climate, and to the interplay of such factors.

To understand the soil as a reality in the field one must first have as complete a knowledge as is possible of the factors of the environment formerly or presently acting in soil formation. They are of prime importance in that they influence the rate and type of processes which differentiate the soil into a sequence of horizons. Having accounted for the variants within the zonal groups, the next problem is to give such profiles a short but meaningful name and, finally, to realize that the profile is not unique, but is related to its neighbours as well as to other similar profiles in distant parts of the world.

1 D. H. Yaalon, TICSS, 7, 1960, V. 16, 119–23

2 H. Jenny, *Factors of Soil Formation. A System of Quantitative Pedology.* McGraw Hill, New York, 1941, 281

3 R. L. Crocker and J. Major, J. Ecol., 43, 1955, 427–8

4 J. S. Bunt and A. D. Rovira, JSS, 6, 1955, 119–28

5 E. A. Flint and J. D. Stout, Nature, 188, 1960, 767–8

6 R. A. Webley et al., JSS, 14 1, 1963, 102–12

7 B. B. Polynov, Doklady (Pedology), 7, 1945, 327–39

8 V. O. Targullan, Poch., 11 1959, 37–48

9 E. Winters and R. W. Simonson, The Subsoil, *Adv. Agron.*, 3, 1951, 1–92

10 D. F. Ball, Nature, 189, 1961, 688–80

11 J. D. Thorpe and G. D. Smith, SS, 67, 1949, 117–26

12 R. W. Simonson, PSSSA, 23, 2, 1959, 152–6

13 A. A. Rode, *The Soil Forming Process and Soil Formation*, transl. J. S. Joffe, IPST, 1962, 1–143

14 S. A. Harris, TICSS, 7, 1960, V. 20, 138–51

2

Factors of soil formation

DOKUCHAYEV (1846-1903) is credited[1] with the idea of a soil as a constantly changing function of five major soil-forming factors— (1) 'of *local climate*, especially of water, temperature, oxygen and carbon dioxide'; (2) of *parent materials* ('mother rocks'); (3) of plant and animal *organisms*, 'especially of the lower order'; (4) of *relief and elevation*; lastly (5) of *age of the country*. But a recent study of the work of Lomonosov (1711-1765)[2] shows that he, too, recognized the importance of external and internal agents in soil formation. To Dokuchayev, however, must go credit for adding relief; of illustrating the prime importance of the climatic factor in soil formation[3]; and greater credit for considering the soil as an independent entity rather than as a mere result of rock breakdown by climate and organisms—or even as an intermingling of minerals by local or distant transport as Morton[4] suggested in 1838 in a work on the soils of Britain.

Most students know of the approach to the soil-forming factors by Jenny[5] who formulated five independent variables—cl'; o'; r'; p; t. The comma denotes internal soil climate, soil organisms, and the shape of the soil surface rather than external climate, vegetation and landform. The two 'internal' factors are parent material and time. For any combinatio of these factors the state of the soil system is fixed and one can equate s=f' (cl' o' r' p, t), though only for a single soil property (eg pH, porosity or clay content), not for the total soil, nor for a component horizon. The time period involved is very short, only a matter of hours or days if moisture and temperature changes are considered. This approach is most applicable in agronomy and soil chemistry.

Jenny also formulated an environmental equation of 'formative and variable' factors—s=f(cl, o, r, p, t....)—which is of greater interest to geographers. In this formula the symbols stand for (1) external climate; (2) all organisms, perhaps including man; (3) major relief form (rather than slope facet, r'); (4) parent material and (5) time, now the total time elapsed since 'time zero'—the initiation of a new surface.

The possibility of other factors being recognized in future studies is implied in this open equation, though suggestions for other factors have not been very convincing.[6] A case can be made out for the activities of man, constructive or destructive of soil, both directly and indirectly. Long-cultivated soils and those developed on pre-historic sites could thus be studied. Nor can one conceive of *time* acting independently of *space*, though most of the effects of contrasting space relationships are better expressed within the existing equation rather than adding another, almost dimensionless, factor. An aspect of space often omitted from otherwise admirable descriptions of soils is their precise location. This can be achieved in latitude and longitude.[7] Gravity, like space, is taken for granted and is obviously a passive factor. Division of internal climate into soil climate and groundwater regime in the first equation, and the addition of external zonal climate and micro-climate in the second is also possible. Crocker proposed an equation s= ∫ f(c, o, r, w, p)t with 'w' as 'water table' (which could only be a dependent factor) and because he thought that mathematically it was more sound to equate soil as an integral of five factors against time.[8] All these additions, debasing the simplicity and integrity of the equation, at least clarify its scope and purpose.

It is by no means clear whether the factors in the environmental equation are placed in decreasing order of importance or magnitude.[9] The most complex factor is placed first and none of the others is free of its influence. The range of climate limits the type of organism and landform which may develop; it also controls the rate of formation of parent material. Climate admittedly may change with time. The plant cover, by succession, changes its form without change of climate and creates a new micro-climate at the soil surface in so doing. The last two factors, parent material and time, are those most capable of direct and accurate quantitative assessment.

The climatic factor has many components both independent and dependent, and theoretically their variations are measurable. Organisms can be dealt with systematically by counting populations and plant growth rates—the biomass of quantitative biology—and by considering the zonal growth conditions of plants and fauna. Though the interaction of soil and any one type of organism can be assessed, the total effect of all organisms is hard to infer with any degree of accuracy. These organisms are also continuously interacting with each other, usually in predatory fashion. Another difficulty with the organic factor is that it is represented by, and operates through, both living and dead material of both plant and animal origin.

Local and regional effects of relief may be measured and analysed with some precision by the techniques of quantitative geomorphological analysis. Parent material may be characterized by comparative physical, chemical and mineralogical analyses of rocks and soil horizons to show the course and sequence of weathering. Time would seem to present fewer difficulties, for soil profiles may be dated by geomorphic, palynological or historical dates or else by C-14 datings,[10] always provided that suitable profiles can be found!

At various times in the development of soil science one or other factor has been emphasized as the most significant in soil formation. Before Dokuchayev wrote, emphasis on the texture and chemical composition of soil implied an acceptance of parent material as the dominant factor, an approach which is all too common in geographical texts even now. Dokuchayev and his followers stressed the climatic factor, though one of them, Sibirtsev (1860-1899), stated that 'only parent material could result from weathering and not soil, for biological agents are necessary for distinct horizons to form'.[11] The Russian biologist, Ponomareva, claims that the biological factor is the only effectual factor in soil initiation—the others forming an environment in which the biological factor can function.[12] Thus to some Russian scientists the soil is a 'surface geobiological formation', or 'biogeocœnose', and time 'has no effect but only governs the amount or extent of the operation of other factors'.

In Western Europe, Australia and North America more stress is laid on the physico-chemical aspects of parent material, and on the time factor in soil genesis. Carroll writes: 'the factor of greatest importance . . . is the amount and pH of water which passes through the soil and then, interacting upon the parent material, is responsible for mineralogical changes there and within the mineral and organic complex which is the soil'.[13]

The factorial approach has great appeal for geographers, for all six factors are important parts of the study of physical, plant and agricultural geography. Yet there are serious drawbacks to its use, not least of which is in Gerassimov's challenge that the equation has never been solved—at least for the external factors.[14] Yet this objection takes the equation too literally, considering it an end rather than a means to an end. For the first equation the solution is simply expressed as a soil characteristic. Data for both internal and external factors could be included and the resulting soil 'quality' —soil productivity—could be expressed by the yield of a crop. For the second equation, one might be able to deduce the type of soil profile formed if one knew all the factors. Logically, too, one could obtain an impression of the effect of one factor by deduction if the soil type and all other factors were known. Attempts have been made to maintain four factors constant and then assess the effect of one varying factor acting within a known time period.

Probably the greatest difficulty with the equation is that the factors can be other than formative in their effect on the soil profile. There are many influences at work in this way—soil mixing by fauna, freeze-thaw, soil creep and the expansion of clays—so that 'a given soil profile may represent a steady state conditioned by the interplay of propedisotropic factors (those which differentiate the soil into horizons) and propedanisotropic factors, which mix a part or the whole of the profile'.[15] Soil may also be completely eroded if the factors act destructively.

The final difficulty in using the factorial approach is that, though we may gain great insight into the factors, we learn very little about the soil itself—the prime object of our study.

Descriptions of soil and of its enclosing landscape are the essence of the physical geography of soil, so it is best to deal first with these factors individually as components of that landscape, using quantitative data as far as is possible. As soils form through the influence of organisms on rocks and develop into distinct groups under differing climates, I shall reverse the order of Jenny's equation and consider parent materials (the inorganic factor) first, then organisms and climate, before considering relief and time.

1 G. F. Kir'yanov, 'The Philosophical Foundations of V. V. Doku-chayev's work' Sov. SS, 10, 1960, 1034–40, and E. Ehwald, 'A. von Humboldt und V. V. Dokučayev', A. Thaer Arkiv, 4, 8, 1960, 561–82

2 N. P. Remezov, 'M. V. Lomonosov and the Science of Soil', Doklady (Pedology), 3, 1961, transl. Jan. 1963, 39–41. See also G. V. Dobrovol'skiy, Sov. SS, 10, 1961, 1051–6

3 V. V. Dokuchayev, *Russkiy Chernozem*, (St Petersburg, 1883) in *Selected Writings*, Moscow, 1954, 149–86

4 J. Morton, *On the Nature and Property of Soils*, London, 1838

5 H. Jenny, *Factors of Soil Formation*, 1941, p. 15–16

6 A. A. Rode, Poch., 1946, 400–1 and Poch., 9, 1958, 29–38

7 K. D. Glinka, *The Great Soil Groups of the World and their Development* (1919), trans. (1927) by C. F. Marbut, Edwards, Ann Arbor, 1937, p. 9, col. 2

8 R. L. Crocker, Quart. Rev. Biol., 27, 2, 1952, 139–68

9 H. Jenny, PSSSA, 25, 5, 1961, 385–8

10 W. H. Schotte and D. Kirkham, Pédologie, VII, 1957, 316–23

11 N. M. Sibirtsev, *Collected Works*, vol. 2, Moscow, 1953, p. 306. Reference in Sov. SS, 1960, p. 693

12 V. V. Ponomareva, Vestn. Leningrad gos. Univ. 140, 1950. Also in Poch., 9, 1958, 48–56

13 D. Carroll, in Chapters III and VII of H. B. Milner, *Sedimentary Petrography*, Allen and Unwin, 1962

14 I. P. Gerassimov, Poch., 1947, 193–6

15 F. D. Hole, SS, 91, 6, 1961, 375–7

3

The inorganic factor in soil formation

THE soil provides man with all the chemical requirements of life except oxygen. Yet paradoxically *oxygen* comprises 93·8% by volume and 46·6% by weight of the crustal rocks of the earth, which in turn are the source of the inorganic elements ¡n soils.

Of the 14 most abundant elements in rocks, *silicon* (Si) comes next, less than 1% by volume and 27·7% by weight. When magma is cooled, silicon in excess of that required to form silicate minerals appears as quartz (SiO_2) in igneous rocks. Silicate minerals are built up of units of SiO_4^-, in which a cation of silicon is centred in a tetrahedron of four oxygen atoms. Silicates are classified according to the way these tetrahedron occur; or as they are linked together by other cations, for example, by Mg in *olivine*—Mg_2SiO_4.[1,2,3]

Aluminium (Al) forms 8·13% by weight of the crust and occurs in such rock-forming silicates as micas and felspars as well as in clay minerals in the soil. So does *iron* (Fe), 5% by weight of the crust but only 0·43% by volume. The three elements so far mentioned each combine with oxygen to form oxides and are widely present in both soils and rocks. The next group, of four elements, are the bases; *calcium* (Ca; 3·63% w, 1·03% v); *sodium* (Na; 2·83% w, 1·32% v); *potassium* (K; 2·59% w, 1·83% v) and *magnesium* (Mg; 2·09% w, 0·29% v). These elements occur in rock-forming *felspars*, for example in $KAlSi_3O_8$—orthoclase felspar.[3] If the magma is rich in alkalies and in Si and Al such felspars will form, along with quartz and muscovite mica. But if the magma is rich in Ca, Mg and Fe then *ferromagnesian silicates* form, such as hornblende and pyroxene.

The eight elements so far considered form 98·6% of the crust by weight and virtually 100% of the volume of crustal rocks and of the total number of atoms in such rocks. Other elements present in rocks are *titanium* (Ti; 0·44% w), significant in tropical soils; *hydrogen* (0·14%); *phosphorus* (P; 0·12%) and *manganese* (0·1%). Of the other seventy or more elements found in rocks only a few are important in soil. All these figures refer to igneous rocks; the figures for sedimentary rocks differ from these averages, and those for soils even more so.

The listed metallic and non-metallic *elements* are combined in mineral form, either as simple compounds or else more complex substances. *Minerals* result from the cooling of magma, giving crystals of chemical compounds, each of fixed shape, properties and composition. *Rocks* are mixtures of minerals; for example, granite is a mixture of at least three minerals—quartz, mica and felspar. The chemical composition of rock-forming minerals closely reflects the relative abundance of the elements in soils and a knowledge of their formulae is a prerequisite to an understanding of weathering and of soil formation.

Quartz is the most common mineral in rocks and soils. It is insoluble in most natural acids, and forms the resistant and inert skeletal residue of strongly weathered soils—though it can occur at greatly varying particle size, from fine gravel to fine clay. *Calcite* ($CaCO_3$) is another simple mineral but of secondary, not primary, origin. It is soft and readily reacts with weak acids. *Dolomite* ($CaMg(CO_3)_2$) is slightly harder and does not react so readily. *Gypsum* ($CaSO_4.2H_2O$) is one of the softest of minerals; it is readily soluble in water and occurs mainly in desert soils. *Hematite* (Fe_2O_3) is common in soils derived from ferromagnesian rocks and generally the type—or the degree of oxidation and hydration—of iron oxides present in soils reflects the internal soil climate.

More complex and more important rock-forming minerals are the aluminosilicates, mainly divisible into two groups: *felspars* and *micas*. Potash felspar (*orthoclase*) is common in granites and gneiss and is quite hard, yet weathers easily in the presence of water:

$$2KAlSi_3O_8+11H_2O \rightarrow Al_2Si_2O_5(OH)_4+4H_4SiO_4+2K^++2OH^-$$

The right-hand side of this equation represents kaolinite, the re-synthesized 'wreakage' of the highly-weathered primary mineral, plus the colloidel silica and soluble salts. The rate of this reaction is increased in a more acid soil, with H-ions derived from dissociation of water, or from organic or carbonic acids.[4] The lime-soda felspars are very easily weathered (*plagioclase*, $CaNaAlSi_3O_8$), as are the soda felspars, for example *albite*, common in basalt. *Micas* are relatively soft, especially white or *muscovite* mica. Its composition— $H_2KAl_3Si_3O_{12}$—indicates its hydroxyl character; it does not readily weather chemically and on hydration it retains positive ions to form the clay mineral *illite*. Yet physically it is weak and readily disintegrates into platey fragments. Black mica (*biotite*)—$K(Mg,Fe)_3$ $(AlSi_3)$ $O_{10}(OH,F)_2$—is rich in iron, weathers readily, and is a source of plant nutrients.

Other silicate minerals of interest are *glauconite*, an hydrated silicate of Fe and K; and *olivine*, present in volcanic and basic igneous rocks, which form the parent materials of dark-coloured inorganic soils.

Rocks are consolidated aggregates of minerals and occur *in situ* at the base of the soil or as fragments within soil horizons, usually decreasing in size towards the surface. Unconsolidated parent materials may be aggregates of rock fragments or of mineral particles, both in a weathered or in an unweathered state. Rocks are conventionally divided into three groups—igneous (magmatic), sedimentary and metamorphic. *Igneous rocks* may be derived from *intrusive magma* with slow cooling at great depth; in which case large crystals are produced for the constituent ions then have enough time to arrange themselves into the unit cells of the crystals. *Extruded magma* may undergo rapid cooling at or near the surface, producing fine-textured or glassy types, for the atoms have not had sufficient time to arrange themselves into crystal form.

Most minerals in deep-seated igneous rocks are not affected by weathering before crystallization and are very unstable in contact with the atmosphere. Their weathered residues are very variable, either coarse or fine in texture depending on the depth of crystallization, and are either base-rich or acidic. Igneous rocks may be classified according to texture, which is useful in comparing their texture with that of the soil, and is a property which influences their weatherability; or they may be classified according to their content of silica. Igneous rocks with an SiO_2 content $>66\%$, such as pegmatite and granite, are termed *acid*; rocks such as diorite, with 52%–66% SiO_2, are termed *intermediate*; those with between 45%

and 52%, *basic* (dolerite and gabbro); and *ultra-basic* rocks, dominated by ferromagnesian minerals, have $<45\%$ SiO_2.

Igneous rocks are rather rare as parent materials for the developed soils of cool temperate latitudes and tundra areas, except in eastern Scandinavia and eastern Canada. Usually they have provided source materials for glacial, alluvial and colluvial parent materials in upland western Europe. They are perhaps more important in mountain areas, where only azonal soils are formed on thin transported debris layers, and achieve greater importance as parent materials in the old shield areas of Africa, South America, Australia and the Piedmont of south-east USA.[5,6] In Britain they are found in the south-west (granite), in Northern Ireland (basalt), in Scotland and in the Malverns.[7] They occur widely in central France—in, for example, the granite of the Auvergne.[8] Studies of the soils in the Malverns show that the primary minerals of igneous rocks weather to distinct types of clay minerals in the soil. In some granites, felspars have been partially decomposed at depth by hydrothermal reactions and this decreases the durability of the surface rock.[8,9]

Sedimentary rocks have been derived from igneous rocks during past periods of erosion and deposition. Their composition and particle size are related to the properties and weathering history of the igneous rocks. Globally they are classified according to their degree of consolidation into (1) diagenic or cemented deposits, hardened into true rocks with particle size decreasing in the order: breccias, conglomerates, gritstones, sandstones, shales; (2) biogenic sediments such as limestone and dolomite; (3) chemical precipitates such as gypsum and (4) unconsolidated sediments. This fourth class is best subdivided according to mode of formation: (a) wind (loess, dunes); (b) water (alluvial, lacustrine, fluvioglacial and marine sediments); (c) (i) ice action (till), (ii) periglacial action (sludge deposits); (d) gravity (screes); (e) volcanic ash and tephra (dust) and (f) organic accumulations such as peat, fen, bog, moss and lichen. There are also those of complex origin, for example, (g) clay-with-flints[10] and colluvial deposits.

Most sedimentary rocks are well sorted by density or particle size into well-bedded aggregates with distinct geological horizonation. In some the cementing material is dominantly of a carbonate nature, in others it is siliceous or ferruginous and more resistant to decomposition. The carbonate content of rocks and parent materials is important in weathering studies for its reaction to water is closely linked to temperature regimes and carbon dioxide pressures. Chalk contains 95%–100% $CaCO_3$; chalk marls have 60%–85% and chalky boulder clays from 40%–60%.[11] Loess and marine muds usually have 10%–25% $CaCO_3$, but can reach 38%.[12]

One may distinguish two groups of recrystallized metamorphic rocks, *orthometamorphics* derived from igneous rocks and *parametamorphics* derived from sediments. Of igneous rocks granite forms gneiss; dolerite forms hornblende-schist and olivine-rich rocks form serpentine.[13] Compaction of clays or shales produces micaceous schists or slate; quartzite forms from sandstones and phyllites from grits. Hornblende, mica, felspar and quartz are common in metamorphic rocks and certain minerals are virtually unique to them—the hydrous silicates such as *chlorite* and *epidote* $(Ca_2(Al,Fe)_3(SiO_4)_3(OH))$, for example, are both important in certain soils.

Metamorphic rocks are common in mountain regions and in the areas of the older Hercynian and Caledonian orogenies. They are also present in the recently deglaciated shields of eastern Fennoscandia and north-east Canada where soil development has hardly begun on the solid rocks for metamorphic rocks are weatherable only with difficulty and quartzite is perhaps most resistant of all. The foliation of gneiss aids its physical disintegration while micaceous schists are easily hydrated.

Clearly, not all rocks weather with equal ease. The more complex their mineral composition the greater their potential solubility, the more seats of weathering there are on the crystal faces; also there is greater chance of physical disintegration if they are varied in composition. Calcareous rocks may lose up to 90% of their mass on weathering, though the low mobility of organic colloids in Ca-rich material and the minimal contents of non-calcareous residue do not produce soils of great thickness.[14] Igneous rocks containing Ca minerals and Fe oxides weather very easily.[15] Sedimentary siliceous rocks, having passed through one or more cycles of weathering and erosion, rarely lose >60% of their mass, often much less. Surprisingly, clayey rocks and fine-grained sediments are the most common parent materials, far more widespread than igneous, sandy, or lime-rich rocks. Recent estimates show that they underlie from 60–80% of the land surface of the world.

One can scarcely continue this discussion of parent materials without a more detailed knowledge of weathering processes—the chemistry of the interaction of climate and mineral elements—though various examples of the elemental constitution of rocks, substrates, subsoils and soils may be given to illustrate the progressive change in chemical constitution of rock to soil. They also show the loss of mass of the rock in weathering; the relative increase of water in certain horizons; and finally the slight increase of soluble material at the surface derived from plant debris. Table 1 illustrates the loss in a temperate climatic environment:

TABLE 1

CHEMICAL COMPOSITION OF GRANITE, MANTLE AND SOIL, ST GERVAIS, PUY-DE-DOME. FRANCE (46°N)[8]

%weight of dry mineral matter . . . (quartz as % of total mass)

	SiO_2	Al_2O_3	Fe_2O_3	CaO	MgO	K_2O	Na_2O	TiO_2	P_2O_5	H_2O	Quartz
Soil	68·9	14·5	3·6	0·5	1·2	4·8	1·6	0·9	0·43	3·6	31
Sub soil	67·5	15·9	3·8	0·4	1·1	5·0	1·7	0·8	0·21	3·6	28
'Arène'	66·8	17·1	2·5	0·7	1·4	5·1	2·3	0·7	0·23	3·2	23
Rock	67·8	15·6	2·5	1·9	1·6	5·2	3·5	0·6	0·24	1·1	25

Here contents of Si, Fe, Ti and P, as well as water and quartz, apparently increase from rock to soil; Ca, Na, Mg and K decrease. However these relative changes may not be shown if only the colloidal mineral matter is analysed in the three altered layers, and the behaviour of Al and Fe in the uppermost layers may vary, especially the latter, for the mobility of iron in Fe^{++} and Fe^{+++} form is very variable. Nevertheless, as a global average for igneous rocks, one may formulate a list of the increasing relative mobility of elements:

$$Ti < Al = Fe < Nb < Si < K < Mg < Na < Ca < S < Cl$$

and the last three elements are extremely mobile in comparison with the others (Fig. 3).

For purposes of comparison with table 1, here are analyses of soils derived from basalt and granite under tropical conditions:[16]

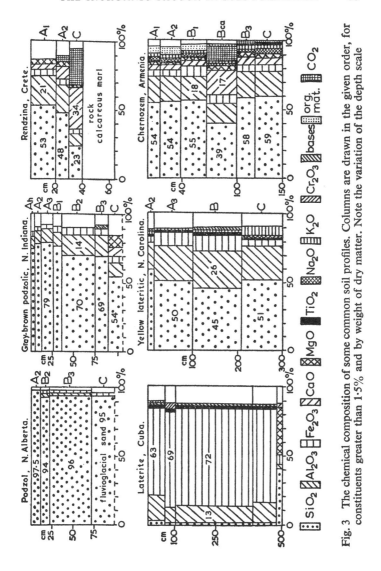

Fig. 3   The chemical composition of some common soil profiles. Columns are drawn in the given order, for constituents greater than 1·5% and by weight of dry matter. Note the variation of the depth scale

B

TABLE 2

COMPOSITION OF BASALT AND SOIL LAYERS, KAUI, HAWAII (22°N)

(% WEIGHT)

	SiO_2	Al_2O_3	Fe_2O_3	CaO	MgO	K_2O	Na_2O	TiO_2	P_2O_5	H_2O
Soil	9·2	24·4	35·8	0·3	0·3	0·21	—	6·89	0·37	15·0
Subsoil	9·9	28·9	35·4	0·2	0·2	0·06	—	5·54	0·40	17·1
Mantle	32·8	24·0	21·0	3·8	2·4	0·21	0·34	3·54	0·45	10·3
Rock	49·0	13·7	13·2	7·3	13·5	0·27	1·62	1·73	0·13	0·4

In this sequence Ca, Mg and Na are completely lost from the rock and Si also decreases greatly; Fe, Al, Ti, P and water (as well as organic matter) have increased in the soil. Plants have depleted Si, P and water in the surface but have returned K, Mg and Ca to the topsoil and P to the subsoil and mantle. However, the soil has a bulk density only one-third of that of the original basalt. If one assumes that the mineral matter in the topsoil is exclusively derived from the rock, then each figure should be multiplied by the bulk density to obtain the content of material in grams per cc before a true comparison could be made of both the relative amount of loss or gain between one horizon and the next, as well as additions and extractions by plants.

Aluminium is often regarded as a static constituent of decomposed rocks and derived soils, having low mobility and existing in a fixed form. It can therefore be used to obtain a minimum figure for the relative losses per unit volume of material. A comparison of % weight alteration of granite and of % volume changes based on Harrison's classic data of katamorphism in a tropical environment has been given by Lovering:[16]

TABLE 3

COMPOSITION OF GRANITE AND SOIL LAYERS, MAZARUNI,

BRITISH GUIANA (6°N) (% WEIGHT)

	SiO_2	Al_2O_3	Fe_2O_3	FeO	CaO	MgO	K_2O	Na_2O	TiO_2	H_2O	Quartz
Soil	69·7	23·2	1·6	0·05	0·01	0·17	0·5	0·1	0·94	4·7	39·9
Subsoil	62·4	24·4	1·5	0·27	0·02	0·22	1·4	0·1	0·84	8·6	31·5
Kaolinized Arène	70·2	18·8	1·8	0·25	0·03	0·17	2·7	0·2	0·68	5·2	41·7
Rock	72·7	14·5	1·4	0·50	1·03	0·82	4·8	2·8	0·67	0·7	31·7

In this table soluble silica may be obtained by subtraction (SiO_2 − quartz). Thus soluble silica decreases toward the surface (41 % rock; 30 % soil) while quartz increases. Bases decrease toward the surface as does FeO; while Fe_2O_3 shows a slight increase, and Ti and Al marked increases, toward the surface.

TABLE 4

SAME MATERIAL (% VOLUME)—AL CONSTANT

	SiO_2	Al_2O_3	Fe_2O_3	FeO	CaO	MgO	K_2O	Na_2O	TiO_2	H_2O	Quartz
Soil	34·9	11·6	0·8	0·02	tr.	0·08	0·2	0·05	0·47	2·4	19·9
Subsoil	29·6	11·6	0·7	0·13	0·01	0·10	0·6	0·06	0·40	4·1	14·9
Arène	43·2	11·6	1·1	0·15	0·02	0·10	1·7	0·13	0·42	3·2	25·7
Rock	72·7	14·5	1·4	0·50	1·03	0·82	4·8	2·80	0·67	0·7	31·7

Clarification comes from the volume changes, which show that 41% of the original rock is lost during its weathering to clay substrate (arène), after which the Al is almost stable, while Si, Fe and quartz all suffer further loss of volume—even though they are relatively more important by weight. The bases lose most of their volume between rock and 'arène' with the exception of K, whose final loss is somewhat delayed. Only 44% of the original rock is left in the subsoil. The increase of Si at the surface is due to plant decay, while the volume moisture content is greatest in the subsoil —the seat of maximum chemical weathering.

With greater certainty one may assume that the volume content of Ti remains constant during the weathering of rock and the formation of soil, especially if ilmenite ($FeTiO_3$) is present. Losses and gains at Angadipuram, near the type area for laterite, are therefore illustrated with Ti content held constant at 0·63%:[17]

TABLE 5

COMPOSITION OF ROCK, SOIL LAYERS AND LATERITE, ANGADIPURAM, MALABAR (11°N) (% VOLUME)

Depth (m)	Horizon	Combined SiO_2	Al_2O_3	Fe_2O_3	CaO	MgO	K_2O	H_2O	Quartz	Total
0–0·3	Soil	5·6	6·3	7·0	0·16	0·14	0·5	2·7	13·2	35·6
0·3–1·8	Laterite	11·6	10·0	5·6	0·06	0·08	0·1	3·2	10·1	40·6
1·8–5·5	Red clay	10·7	8·6	4·8	0·09	0·22	0·3	3·8	12·0	40·5
5·5–6·7	Mantle	4·1	11·0	8·8	4·54	1·07	4·3	1·3	26·8	61·9
at 7m.	Rock	55·4	21·3	4·6	6·93	1·38	7·3	0·1	—	97·0

These total volume-loss figures show that 35% of the rock is lost in the first stage of weathering to the mantle and the mass further decreases to 41% of the original rock at between 2 and 6 metres depth. The surface soil represents a further loss of 5% and is equivalent to only 36% of the original rock. In a laterite developed on basalt at Kasaragod (S. Kanara), the loss was by 46% to give only 38·6% of the original rock in the mantle and 32% in the clay horizon, but the return of bases by plants to the surface soil was greater on the basic than on the acid rock.

All rocks, therefore, are subject to gains and losses in the process of their transformation to mantle and soil. Often the loss is very great in the first stages and the final net effect in the upper soil represents a considerable but not greater loss, for the bulk is often increased by the addition of organically-derived debris, by water and by air. As water, through hydration, hydrolysis, solution and transport, is the chief means of breaking down rocks and their component minerals to produce fine alteration products, so, in general, 'the factor of parent material increases in importance in determining soil type as one passes from wetter to drier climates'.[18] One may add that the effect of parent material decreases with the increasing passage of time from a 'time zero'.

1 F. W. Clarke, *Data of Geochemistry*, USGS, Bull. 770, 1924

2 B. Mason, *Principles of Geochemistry*, Wiley, NY, 1952; and V. Goldschmidt, *Geochemistry*, transl. A. Muir. On silica in soils see J. A. McKeague and M. G. Cline, *Adv. Agron.* 15, 1963, 339–96

3 *Rutley's Elements of Mineralogy*, 24th ed., by H. H. Read, Murby, 1960

4 D. H. Yaalon, J. Chem. Educ., 36, 1959, 73–6

5 L. Berry and B. P. Ruxton, JSS, 10, 1, 1959, 54–63

6 F. B. Alexander, QJGS, 115, 1959, 123–44. Also J. G. Cady, PSSSA, 15, 1950, 337–42

7 I. Stephens, JSS, 3, 1952, pp. 20–33 and 219–237

8 D. Collier, Ann. Agron., 12, 3, 1961, 273–331

9 G. P. Merrill, *A Treatise on Rocks, Rock-Weathering and Soils*, MacMillan, London, 1904

10 J. Loveday, PGA, 73, 1, 1962, 83–102

11 C. H. Bornebusch and K. Milthers, *Soil Map of Denmark*, Dan. Geol. Undersøg. Rk. III, nr. 24, 1935

12 F. Kohl, ZPDB, 75, 1956, 114–131; B. Verhoeven, Int. Inst. Land Reclam. Bull. 4, Wageningen, 1963

13 J. R. Butler, Geochim. Cosmochim. Acta, 4, 1953, 157–78

14 C. D. Piggott, J. Ecol. 50, 1962, 145–56

15 R. Glentworth, Tr. Ryl. Soc. Edin. LXI, I, 5, 1944, p. 162; and P. Schauffelberger, Schweiz. Min. u. Pet. Mitt. 32, 2, 1954, 319–35

16 T. S. Lovering, BGSA, 70, 1959, 781–800

17 K. V. S. Satyanarayana et al., J. Indian SS, 9, 1961, 107–18

18 G. Milne, A Soil Reconnaissance Journey through parts of Tanganyika Territory, J. Ecol. 35, 1947, 196–265

4

The organic factor in soil formation

SOME geographers do not care to regard the soil as a biological phenomenon, in large part only superficially altered by man. Yet the close integration of most life forms in soil formation from the very earliest stages must be admitted. Other geographers regard the soil as a part of the study of 'biogeography', rather than as an independent study; then neglect the pedobiological aspects, which is an even more regrettable state of affairs. The purpose of this section is to enumerate the form and amount of plant and animal life acting in or on the soil, and to characterize the form and role of the dead organic matter within it.

The amount of solar energy used by plants in CO_2—assimilation is only 0.01% of the total solar energy reaching the earth. By this means light energy is transformed into chemical energy and the elements in the atmosphere are brought into the soil. The basis of all life on the earth, therefore,[1] is the process:

$$6CO_2 + 6H_2O + (708 \text{ kg/cal. energy}) = C_6H_{12}O_6 + 6O_2$$

The sugars produced form cellulose and other complex substances of the plant dry matter and, by respiration, plants extend their roots by oxidizing carbohydrates—$COOH + O_2 \rightarrow CO_2 + H_2O +$ energy—to obtain nitrates and mineral nutrients from the soil. Living plants and animals of all sizes achieve mechanical and bio-chemical effects by this energy transfer. On death they deliver a considerable amount of material to the soil which has four effects: it helps sustain other forms of life, changes the appearance of the surface soil, accelerates the weathering of minerals and promotes soil formation processes.

Plants provide debris to the surface which is broken down on digestion by lower forms of animal life. Among these are bacteria, protozoa, insects and worms.[2] In many soils much of the organic material is in the form of coprogenous faeces and slime from their activity.[3] Plant debris itself consists of leaves, needles, twigs and fruits as well as roots which, on decay, are a source of organic matter at greater depth in the soil additional to their physical and biochemical effects when alive.

All plant debris is composed of *inorganic* and *organic* substances. Of the inorganic elements, lime, K, Mg, Fe, and others such as P and S, are found, which remain as *ash* when all organic material has been removed. The *organic* portion of the debris consists of *non-nitrogenous compounds* such as starches, sugars and lignin and complex *nitrogenous substances* such as amino-acids. The first are made up of sugars or carbohydrate molecules (COOH), which combine with oxygen to form cellulose, and other substances which can be split up on decomposition. Usually cellulose loses 75% of its mass on mineralization to CO_2 and H_2O and only 25% is synthesized into the tissue of bacterial or other organic forms.[4] Amino-acids contain C, O, H, N and S, and nucleoproteins contain phosphoric acid. Also present as decomposition products are complex *organic acids*, resins and oils composed of various combinations of C, O and H.

To gain a first impression of micro-organic activity we may refer to the ability of lichens to thrive on apparently bare rock surfaces, holding a film of water near to them, extracting nutrients from rock minerals by ion exchange, and obtaining nitrogen from the air. These sheets of lichen die off rapidly forming a 'protolitter' which provides support for other vegetable matter. This is not really a soil until considerable bacterial humification has occurred and some fine earth has accumulated by the breakdown of minerals. In addition, in producing CO_2, bacteria aid the further breakdown of rock minerals by carbonation as well as by chelation.[5]

Microflora
Bacteria are the most remarkable of all soil organisms. Being plants with but a single cell, they are smaller than the smallest mineral soil particles, varying from the smallest, 0·001 mm in length, to the largest with 0·005 mm ø. It is almost impossible to conceive or to calculate how many exist in the soil, though they are perhaps

the most numerous of all soil organisms.[2] Recent estimates vary between 1 million and 4000 million per gram of soil.

Their activity within the soil determines their grouping into (1) *autotrophic bacteria*, which oxidize mineral substances such as sulphur and iron for their metabolism, gaining carbon from CO_2 and N from inorganic compounds, and (2) *heterotrophic bacteria* which obtain energy solely from organic substances. Some of the latter fix nitrogen and others require it. Of the N-fixers some are dependent on legumes and are known as *symbiotic bacteria*; others are *non-symbiotic*, existing independently, such as azotobacter in aerobic surroundings and clostridium in anaerobic conditions. *Azotobacter* are widely found in soils of neutral or alkaline reaction (pH$>$5·8) and need lime-rich carbonaceous organic matter.[6] These relatively large bacteria are not numerous in most soils (100–1000 per gram) and do not develop well in soils of low humus content or where soil temperatures are low, though drought is even more damaging than frost. In the arctic they are absent and autotrophic bacteria disintegrate the rock. *Clostridium* are acid-tolerant, more numerous and widespread, with up to 100,000 per gram of soil. A third non-symbiotic N-fixer is *Beijerinckia*, which is acid-tolerant and confined to tropical and monsoon lands. All these organisms fix from 6 to 55 kg/ha/yr N, which cannot compare with the work of the symbiotic N-fixers (340 kg/ha/yr).[7]

Of the *symbiotic bacteria*, some form spores, others do not, and both have aerobic and anaerobic types. Apart from the well known nodule formation on legumes, accomplished by *Rhizobium*, other N-fixers act on alder trees or marsh plants. Of the symbiotic bacteria most are tropical in origin and the much studied species of temperate lands, developed from these, are adapted to high nutrient, lime-rich, soils.[4]

Most bacteria readily accomplish the breakdown of the complex carbon compounds of plant debris by unknown processes and the role of any one type is hard to establish.[2] Much carbohydrate is converted to CO_2 and water during debris decomposition. Some of the N of the organic matter is set free as *ammonia* (NH_3) while the N actually ingested by organisms is later excreted in similar form. Aerobic bacteria form enzymes which rapidly oxidize *ammonia*, especially in moist soils, to form *nitrite*: ($2NH_3+3O_2=$

$2HNO_2+2H_2O+$energy), and then *nitrate* on further oxidation: ($2HNO_2+O_2=2HNO_3+$energy). Nitrite is present in minimal amounts in the soil and is rapidly transformed into nitrate, which is highly soluble, varying from 2 to 20 ppm in most soils, reaching 100 ppm in horticultural soils.[4]

There is wide seasonal variation in the rates of *ammonification* and *nitrification*. Both are most intense in the upper parts of the soil—in summer in cool regions and throughout the year in warmer climates. In Denmark NO_3 content in sandy soil in June was 110 kg/ha, with an annual average of 62 kg/ha. The annual net production, after losses, was 40 kg N in the top 60 cm in sands and 60 kg N in loams.[8]

Azotobacter can fix 1 part of N for every 20 parts of C which it transforms,[2] and the *ratio of C to N* in soils is often taken as an index of the state of decomposition of organic matter by micro-organic activity, for the ratio varies only slightly within any one type of organic matter or soil horizon in a particular soil zone.

TABLE 6

REPRESENTATIVE C/N RATIOS FOR SOIL SURFACE HORIZONS

Podzols				Savanna fallow	chernozem	arid brown	serozem
Scotland		Yugoslavia					
mor	A_2	A_2	B_2				
26	18	14	8	14	11	6	4·5

Thus the *C/N ratio* is related to climate[9]; it characterizes the stage of debris decay; and is a partial measure of soil fertility, for the availability of N to plants is also expressed by this value.

There are other factors at work varying the C/N ratio but, in general, the organic matter of mineral soils contains C, N, and organic P in ratios 110:9:1; in peat the ratios are 229:10:0.4. All the ratios are wider in organic soils.[10] C decreases in alkaline and calcareous soils giving low ratios; climatic areas such as north-west Europe, with heavy winter rain, have marked leaching of N and C/N ratios are wide, narrowing as the climate becomes drier.[11] Tropical soils are relatively rich in N, with C/N ratios at the surface of 12–16, of 3–8 at 1·5 m—each decreasing in the dry season.[12]

The number of *fungi* in soil varies from 5000 to 1 million per gram which is numerically less than bacteria, though their mass per unit of soil is greater. Most soil fungi are extremely small and comprise (rare) yeasts, moulds and mildews.[13] All start life as spores and produce thread-like hyphae which lengthen and exert a physical effect by pressing into minute spaces in rocks, as well as complex chemical effects by secreting enzymes to digest the organic matter, producing CO_2 as they do so.

Soil fungi are heterotrophic and their food varies from organic acids and carbohydrates to nitrates, while larger puff fungi are able to feed on lignin. All are efficient feeders, retaining up to a half of the carbon they consume within their cell substance.[2]

Fungi are more sensitive to environmental conditions than bacteria and require a fairly high degree of moisture, though they differ in habitat, some preferring base-rich soils, some acidic peaty soils. They develop best in neutral soils though they are most numerous, and spore most readily, in sands or in acidic organic matter. The seasonal distribution of numbers and dominant species varies. In Britain they are most numerous in the autumn, and gentle falls of rain in warm seasons activate them greatly.[13]

The maximum temperature for fungal growth is 35°C, declining if the surface is hotter than this, or if the soil dries out to wilting point. Fungi flourish at low temperatures and the greatest decline in numbers comes when soils are saturated or flooded, hence their absence from the tundra. In the surface soils of deserts in the USSR fungi numbered 35,000 per gram of soil; in northern coniferous forests 230,000 per gram. Species of *penicillium* amounted to half the total fungi in most Russian soils, only 30% in semi-arid and 21% in desert soils.[14]

A curious microfloral form between fungi and bacteria, the *actinomycetes*, are abundant in soils (<40 m p gm). Their bio-mass is intermediate between bacteria and fungi. Aerobic, able to survive in dry soils, their growth ceases below pH 5. They attack many humus substances beyond the reach of bacteria, and are efficient N-releasers.[15] Best known are *streptomycetes*, though many actinomycetes produce antibiotics which influence the soil micro-fauna, while fungal-produced antibiotics affect only the microflora.

Soil algae are filamentous or single cell organisms which are the precursors of higher plants, being blue-green in neutral soils or green in acid soils. They are chlorophyll bearers, taking up CO_2

from the air and using water and radiant energy to form carbohydrates, and some fix N to form protein. They have a world wide
distribution wherever light, warmth and moisture are present. Moist
soils are rich in numbers and species and they retard the leaching
of soluble nutrients. In fertile cultivated soils of temperate lands
they may number 100,000 per gram in summer and up to 3 m p gm
in warm moist soils in autumn and spring. They are among the first
colonizers of bare rock or eroded soil, creating a growth medium
from CO_2, N and minute amounts of mineral matter.[16]

Lichen is an association of algae and fungi, the latter the chief
providers of nutrients by ion exchange with the rock minerals. On
the death of a layer of lichen, moss and higher plants come in. Rock
colonizers are termed *lithophyllic lichen*, the later forms, growing
with moss, *plastic lichen*.

Microfauna

Protozoa are the simplest form of animal life, larger than bacteria
(5–80 μ in length), though unicellular. A few propel themselves in
the soil water by means of cilia—hairlike projections, 10μ in length
—hence *ciliates*; others by whip-like tails, 20μ in length—*flagellates*;
while the *amœba*, 10–40 μ in size, consist of unconfined protoplasm,
propelling themselves by protrusions of the cell or 'pseudopods'.
Inert protozoa are known as *cysts*.[2] Protozoa occur in most parts
of the world and in one gram of soil there may be one million
flagellates, a quarter of a million amœba and 1000 ciliates, with a
weight of 170–335 kg/ha. Most feed on bacteria, especially azotobacter, and they stimulate the N-fixers.

These micro-organisms, both fauna and flora, show intense life
in one gram of soil—billions of micro-organisms. Yet Russell[2]
reminds us that they form only 2–3 % of the total organic matter in
one gram of soil. Their importance is in reducing organic debris
to its elemental constituents, promoting mineral ion exchange and
making nitrogen available to plants.

The rest of the soil fauna falls into two groups, *meso-fauna* and
macro-fauna;[17] meso-fauna are too small physically to disturb the
soil and live in the water in the soil pores; macro-fauna make their
own passage-ways and are capable of feeding on undecomposed
plant litter.[18]

Meso-fauna. Among the worm-like animals in soil are *nematodes* or 'eelworms'. Billions are found in each acre of soil and they are most numerous in grasslands (20 m per sq m) and in conifer plantations (40–250 kg/ha). Most species concentrate near the surface and are invisible to the naked eye, being <1 mm long and 0·02 mm ø. They cannot move into the finest pores of the soil and inhabit the soil water, improving aeration and mixing mineral and organic matter. Most feed on decaying organic matter and the attached fungi; some are harmful, some beneficial—most harmful are those infesting crop roots.[4]

Other minute worms are *enchytraeids*, five to ten times the size of nematodes. They live mainly in acid humus soils, in Danish heath 'mor' reaching 200,000 per sq m.[20] When the land is cultivated their numbers fall sharply to 10,000 per sq m.

Mites (Acarina) are common in spruce mor (100 kg/ha) and reach 80% of the total soil fauna, their chief diet being fungi. *Collembola* (the insect, springtail) may reach 700 kg/ha in beech mull. Spiders, slugs, beetles and grubs all play their part in woodland and grassland soils. *Myriapods* (centipedes and millipedes) are numerous, centipedes feeding on insects and young mesofauna, and millipedes, wildly inefficient vegetarians, chew large amounts of litter to obtain a small amount of carbohydrate. *Snails* are common in lime-rich soils, and can wear deep holes in limestone.

The main role of the meso-fauna is in consuming plant debris and its enclosed bacteria and fungi, reducing the humifying material to colloidal dimensions, and moving it more deeply into the soil.

Of the larger *macro-fauna*, woodlice (*isopods*) are common in dry soils and replace earthworms as litter destroyers. *Earthworms* are, however, the main form of soil macro-fauna in all but the driest or most acid soils. All worms cut up and ingest large amounts of plant debris which is humified in their intestinal tract. The humic acids formed are neutralized by lime secretions and excreted as casts. This prepares the debris for consumption by smaller organisms and releases much of its nutrient material.[2]

Worms are the largest and heaviest of the soil fauna. Ponomareva found 2·94 m/ha in the soils of Russian oak forests, 610,000 in spruce forests and 880,000 in wheat fields. The 1·79 million present in one hectare of a two-year ley took up 50% of the soil of the root zone in casts.[21] Worms are most numerous on lime-rich soils; they are rare or absent below pH 4·5, on dry sandy soils, or in anaerobic conditions, and are adversely affected by frost. They are very numerous in the prairie environment, burying deeply to avoid summer drought. The world record is for old lime-rich pasture in New Zealand (8 m/ha) with a weight equal to that of the sheep carried on the same pasture.[22]

Darwin[3] found that worm casts amounted to 10 tons per acre and the covering of casts on the soil implied an annual accumulation of one-fifth of an inch, though this rate decreased with time as the worms redigested their old casts. Ponomareva[21] found that the amount of soil passed through the bodies of cast-making worms approached 10 tons per acre as a mean and 20 tons per acre as a maximum. It enriched the soil surface in neutral humus and fine mineral material of a water-retaining silt and coarse-clay calibre. Casts contained more humus and exchangeable bases than the rest of the soil, had higher pH and more nitrates. Casts showed greater stability in rainy periods than mineral-bound aggregates, thus worms are the main producers of the organo-mineral complexes important in crumb formation.

Only two British species make casts—*Lumbricus* and *Allolobophora*; the more favourable soils have Allolobophora, living down to a depth of *c* 30 cm, while acid soils (>pH 4) have large Lumbricus species, particularly *L. terrestris*, capable of burrowing to a depth of 1·5 m, its channels aiding the penetration of air, water and roots.[2] Thus the effect of worms is not only physical and chemical in breaking down plant debris, but more intensively mechanical than any other faunal form.

Ants also improve soil structure and are most important in tropical soils. *Termites* are the most active and best known, some feeding on wood, some on humus and others cultivating fungi. Termite mounds reach heights of 6 m, and 18 m ø, in the Congo, and are rich in nutrients at first—in a fixed and unavailable form— but become acid with time.[23] Soils under cleared mounds are rich in Ca and Na even in areas dominated by acid ferrallitic soils,[24,25] but are difficult to cultivate for many reasons.

Of the larger animals, the burrowing of *rodents* is localized but causes mixing, aeration and collapse of the soil. Colonies of prairie dogs in Texas occupy areas as large as 40 acres, their burrowing destroying soil structure and hindering the leaching of free lime.[26] The largest animals compact the soil and, in treading paths, destroy the turf cover and so render it prone to erosion.

Birds consume earthworms, the smaller meso-fauna, and tree insects. Their droppings form a supply of nutrients for plants, especially in areas of primary colonization. In Antarctica, penguins are the main source of organic matter. Their excrement, together with blue-green algae and keratin from feathers, provides a distinctive *ornithogenic soil*.[27]

Macroflora. The larger plants create a distinct climatic environment above the soil and bring, into and on to the soil, elements removed from the atmosphere and converted from a gaseous to a solid form. More direct effects of plants are the removal of water from depth, protecting the surface from wind and water erosion; physically and chemically altering mineral material by their roots and, more obvious, providing a layer of litter—the main source of organic matter within the soil on its decomposition by micro-organisms.

In studying soil organic matter four concepts must be kept separate: litter; plant residue; mobile humus; and *organic matter*. Organic matter implies all living and dead matter in and on the soil. *Plant residues* are non-decomposed or partly decomposed material on the surface—Wilde's '*ectohumus*'[28]—the L (litter), F (fermentation) and H (amorphous humified) layers above the mineral soil. It may be described as 'mego-organic' and is usually referred to as *mor*. *Litter* is the non-decomposed freshly fallen leaves, fruits, dead twigs and stalks; also leaves which are partially decomposed by micro-organisms before falling. Finely dispersed and altered organic matter moved into the soil, and interacting with mineral matter is termed *mobile humus* or '*endohumus*'. It is amorphous and consists largely of coprogenous material. There are two fractions of endo-humus: (1) an *adhesive fraction* of high adsorptive capacity and (2) an *ionic fraction*, much less adsorptive and divisible into organic and mineral parts.[28] The adhesive fraction forms up to 90% of the humus of prairie soils, and decreases in favour of ionic in tropical and temperate forest soils in which there is light coloured '*crypto-humus*' as pale yellow organic compounds.

Litter. The fall of leaves and twigs has been measured in various environments. For tropical regions (Colombia and Costa Rica) annual falls of 85–120 hkg/ha compare with 9–31 hkg/ha in the Sierra Nevada of California.[29] The high fall in the tropics is balanced by faster decomposition and a quasi-equilibrium of fall and decay is reached in 10 years, while the plant debris layers continue to thicken for longer periods in cooler latitudes. In Russia, oak yields 45 hkg/ha litter per year, larch 82, spruce 123 and pine 250 hkg/ha/ yr.[30] Defoliation of oak begins in autumn and lasts for 3 months, varying from 2·8 to 5·5 tons/ha. In spring there is a maximum mass of 13·2 tons/ha of L and F layers which declines and thins to a minimum of 9·7 tons/ha in autumn. Under pine in spring there is from 37 to 46 tons/ha. Thus the maximum mass of residue in oak forest is roughly equal to the annual litter fall in a tropical forest, and is only one-third the mass of that in pine forest.

Plant residues. The freshly fallen litter loses half its mass in Russia in the summer months, declining from 5·3 to 2·75 tons/ha. The F layer decreases from 7·9 to 6·95 tons/ha—by 12%—in one summer. It consists of decomposed leaves, dark brown in colour, interwoven with fungal mycelia, and is termed 'F' because of fermentation by micro-flora. The L and F layers are easily depleted of water soluble substances in the order (K, P, N, Ca and Mg) and become richer in Si, Al and Fe.

As the mass decreases by carbohydrate decomposition, so the ash content increases, from 5% to 6·5% in the F_1, and from 8% to 9·4% in the F_2 layer. In Russia weight losses of beech litter were 25% after 1 year, of oak 34%–39%, and of mulberry 64%. Deciduous litter usually decays more quickly than the litter from conifers, and leaching of soluble substances is greater in anaerobic than aerobic litter.[31] In the southern Appalachians broadleaved litter lost 35% in 1 year at 1600 metres and 46% at 260 metres. Conifers there lost 29% and 40% respectively.[32] On average the loss of litter in south Finland was 40% higher than in Lapland and was faster from birch than pine, birch losing 70% after 3 years in the south, and only 52% in Lapland—the rate of litter loss being directly proportional to the mean summer temperature, in the range $+3°$ to 40°C. Plant residues are often quite moist, with an absolute minimum moisture content of 30% in temperate lands[34]; oak litter, for example, retains 20–25 cu m/ha water in its lower parts, aiding decomposition.

Organic Matter. The production of living matter by plants in various climates—the so-called *biomass*—is relevant to soil studies, for this mass largely consists of C, N, O and H, of which a certain portion is annually given up to the soil surface.[35] Recent information (Table 7)[36] shows that plant residues accumulate with increasing age of vegetation, and that tropical soils under gallery forest have large root contents and less surface residue than temperate soils.

TABLE 7

VARIATION OF FOREST BIOMASS WITH AGE AND ZONE (1000 kg/ha)[36]

	Location	Age (yrs)	Number of trees per ha	Plant residue	Dead matter on trees	Roots in soil
Pine	E. England	23	3640	30	13	28
		55	760	45	10	34
Pine	S. Scotland	33	4260	111	19	36
Tropical deciduous	Ghana	50	6252	23	72	25
Tropical evergreen	Congo	18	c 9000	5·6	17	36
Gallery forest	Siam	100	?	3	?	89

The relative proportions of plant residue, dead material on trees, and roots, vary with time and climate. In tropical forests there is more destruction above ground level; leaf fall continues throughout the year and so does its destruction, and roots are, therefore, an important source of organic matter.

Roots. In subhumid hot climates, fine roots decay rapidly after the annual growth period and leave pores and tubes in the soil. In humid tropical areas roots affect the rocks and weathered mantle to great depths. Tropical swamps produce one-fifth of their live matter, and forests up to 65% of their annual biomass production, in root form.[37]

The roots of cultivated plants are the main source of organic matter in arable soils. In Holland,[38] grass and winter wheat produce 26 hkg/ha, spring wheat only 15 hkg/ha, but lucerne may achieve

60 hkg/ha. Usually the depth and amount of root are related to the moisture available to the plant and the presence of aerated pores rather than to the nutrient content of the soil. Apart from adding organic matter to soils, roots alter the soil and soil water adjacent to them for distances of 2–3 mm. This 'rhizosphere' is more acid, for roots secrete H-ions, and has more exchangeable bases than the non-rhizosphere soil. It may be drier and more leached than the normal soil.

The breakdown of plant residues—mobile humus

Organisms usually select the most nutrient-rich leaves for breakdown and the leaves of ash, elm and hazel are favoured in temperate forests. The mesophyll is decomposed first, then the epidermis and small cells, and the veins are preserved the longest.[30] Purely physical changes are brought about by wind and rain,[31] and some organic material will oxidize when exposed to light. But most decomposition involves organisms. Trichoderma decompose cellulose in summer and so do moulds and many other microflora, so lignin increases and cellulose decreases. Mineralization by micro-organisms results in the formation of gases or soluble substances such as CO_2, NO_3, CH_4, NH_3O and in H_2O production. There is an increase of N, C and H in the upper layers in spring and autumn, with decrease of O_2. The converse is true in summer. Simultaneously with this breakdown there is the formation and synthesis of humic substances— the *mobile humus*—which are complex organic compounds, stable and resistant to decomposition.

If conditions are unfavourable to the activity of microflora and fauna on cold, damp, lime-deficient sites, then decomposition is slow, and dark *mor* forms, with brown F layers and black H layers, a greasy compact structure (or fibrous in peat), with bindings of fungal mycelia and an acid reaction. The C/N ratio is >20 and may reach 50.[9,39] A mineral-rich acid humus form is *moder* with a C/N ratio of *c* 20, which is unaggregated and unmixed. *Mull humus* forms wherever biological activity is intense in a mild moist climate, the debris quickly losing its structure, and being intimately mixed with the mineral matter by the larger mesofauna.[40] It is aggregated, friable, and contains many worm casts. It has a C/N ratio <20; *c* 10 in lime-rich mull. Thus development of *mull* and *mor* depends on external climate, soil water regime, mesofaunal activity and the presence of calcium, for if free $CaCO_3$ occurs, mull will develop even on coarse sand. The debris layers in mor are clearly separable into L, F, H horizons; but in mull they are mixed, though Aml1, Aml2 . . . horizons are distinguishable, with varying colour and organic matter content.

If these organic decomposition products are complex the composition of resynthesized organic products and of mobile humus is very complex. Some are soluble, others are insoluble; some contain, others lack, nitrogen. Decaying plant residues and decomposition products may be termed *simple organic substances*, resynthesized materials may be termed *complex humic substances* though most humus has no definite composition and is continually undergoing change.[15] The simple substances form 10–20% of the total *endohumus*, comprising various groups of substances—carbohydrates, hemicellulose, pentose (sugar) and hexoses (glucose), as well as hydrocarbons. Then there are fatty organic acids such as oxalic acid ($COOH_2$) and acetic, lactic, and saccharic acid, all variants of COOH or CHO. Various resins, esters, alcohols, tannic substances and proteins form another group. Lignin and its derivatives, the residue of older vegetation also belong here and are richer in carbon (62%) than cellulose (45%); they contain nitrogen (3%), and are more resistant to decomposition.

The next and larger group of *complex substances* is of true *humic* nature and forms up to 90% of the humus of developed soils[2]. They are usually combined with mineral material, have a high ash content, and have been characterized as insoluble black *humin* and *ulmin*, and soluble *crenic, apocrenic, humic* and *ulmic* acids.[15] These are removed from the soil in the laboratory by using alkali solutions and are *polymers*. *Humic acids* are complex *polysaccharides*; they contain no cellulose, have heavy round molecules, are made up of carboxyl (COOH), phenolic (OH) and alcoholic (CH(OH)) groups. They contain 50%–60% of C, 3%–5% H, and 6% of N, the remainder being oxygen (30%–40%). These acids participate in soil exchange reactions; they have low pH (3·5), and are hydrophilic and very mobile, except in chernozem. Their particle size is very small— 60–100 Å ø in alkali solutions. *Ulmic acid* is a brown acid and is thought to be an intermediate stage in the formation of humic acid from cellulose, with less oxygen and more water than humic acid in its molecular make-up. *Humin* and *ulmin* are the insoluble *polyphenol* forms of humic and ulmic acid. *Crenic* and *apocrenic* acids are richer in oxygen (48%), and poorer in carbon (45%) and nitrogen (2%) than humic acids and have a lower molecular weight. They are highly mobile, soluble in water, and accomplish the decomposition of silicates. Their names are regarded as obsolete by many workers (as well as the names of many other organic acids) and the term *fulvic* (yellow) acid has replaced the two crenic acids. Fulvic acids are common in podzols in the mor. Their pH is c 2·8, and they are responsible for translocating iron, forming soluble products with alkalis, and insoluble products with sesquioxides, which accumulate in the illuvial Bh horizon of podzols.[41]

Usually fulvic acids dominate in newly-formed humic substances and are more acidic than humic acid having more COOH groups. Eventually humic acids come to dominate in the upper soil layers. In oak litter Zonn and Sokolov[42] measured 115 kg/ha humic acids and 285 kg/ha fulvic acids; in pine, 924 and 1710 kg/ha respectively. Thus in pine litter there is relatively more humic acid, while in peat fulvic acid dominates. In tropical evergreen forests humic and fulvic acids are usually equal, though in drier tropical deciduous forests humic acids dominate.

It is now thought, at least in the USSR, that each climatic area, or zonal soil, has a corresponding type of humus with an appropriate relative composition of C, O, H and N.[43]

Organic matter therefore alters its form and composition in many ways. Clearly the organic factor is complex and there are intricate problems to be solved. There is a whole sequence of events from the fundamental process of photosynthesis through to leaf fall. But this is only one death of a thousand deaths that the organic matter is to undergo through the action of micro-organisms, then through alteration by bio-chemical and chemical means to form complex humic compounds; and to follow a molecule of nitrogen, carbon or hydrogen through its cycle is a fascinating study.

1 P. Boysen Jensen, *Causal Plant Geography*, Kgl. Dsk. Vid. Selsk., Biol. Medd. 21, 3, 1949
2 E. J. Russell, *The World of the Soil*, Collins, Fontana Library, 1961

3 C. H. Darwin, *The Formation of Vegetable Mould through the Action of Earthworms*, 1881

4 E. W. Russell, *Soil Conditions and Plant Growth*, 9th ed. Longmans, 1961

5 B. P. Gradusov et al., Poch., 7, 1961, 59–66

6 H. L. Jensen, Tidsskrift for Planteavl, 53, 1950, 622–49

7 F. Scheffer and P. Schachtschabel, *Lehrbuch der Agrikultur-chemie und Bodenkunde, Teil I, Bodenkunde*, Enke, Stuttgart, 1960, 208–20

8 E. Poulson and P. Hansen, Tidsskr. Planteavl, 65, 1962, 206–34

9 A. M. Durasov, Poch., 7, 1961, 29–34

10 C. H. Williams et al., JSS, 11, 1960, 334–46

11 P. Duchaufour, *Précis de Pédologie*, Masson, Paris, 1960

12 J. A. R. Bates, JSS, 11, 1960, 246–56

13 J. S. Waid, Growth of Fungi in Soil, in *Ecology of Soil Fungi* Liverpool UP, 1960, 55–75

14 E. N. Mishutin, Izv. Akad. Nauk., Ser. Biol. 5, 1960, 641–60

15 M. M. Kononova, *Soil Organic Matter*, in transl. Pergamon, 1961, p. 135

16 D. A. Osmond, Sci. Prog. 49, 1961, 310–19

17 V. J. Chapman, *The Algae*, Macmillan, 1962, ch. 14, pt. 2

18 F. Raw, S and F, XXIV, 1, 1961, 1–2

19 C. H. Bornebusch, *The Fauna of Forest Soil*, Forst. forsøg. i Danmark, 11, 1930, 1–224

20 C. O. Nielsen, *Field Studies in Enchitraeidae*, Naturhist. Museum i Aarhus, 1955, 1–58

21 S. I. Ponomareva, Poch., 8, 1950, 476–86

22 A. J. Low, JSS, 6, 1955, 179–99

23 P. R. Hesse, J. Ecol., 43, 1955, 449–61

24 R. L. Pendleton and S. Sharasuvana, SS, 62, 1946, 423–40

25 J. P. Watson, SS, 13, 1, 1962, 46–57

26 USDA, *Soil Survey of Dawson Co., Texas*, Ser. 1957, 6, 1960, p. 48–9

27 E. E. Syroechkovskii, Zool. Zh. 38, 1959, 1770–75

28 S. A. Wilde, *Forest Soils*, Ronald, NY, 1958, 76–86

29 H. Jenny et al. SS, 68, 1949, 419–32; and 66, 1948, 5–28; and 69, 1950, 63–9

30 N. P. Remezov, Poch., 7, 1961, 1–12

31 N. Nykvist, Studia Forestalia Suecica, 3, 1963, 1–31. See also a series of papers in *Oikos*, 1959–62

32 R. E. Shanks and J. S. Olsen, Science, 134, 1961, 194–5

33 P. Mikkola, Oikos, 11, 1, 1960, 161–6

34 H. M. Thamdrup, *Studier over Jydske Heders Økologi*, 1, Acta Jutlandica, XI, 1939, 1–88

35 P. J. Newbould, Sci. Prog. LI, 201, 1963, 91–104

36 J. D. Ovington, *Adv. in Ecol. Res.*, 1, 1962, 103–92

37 Roots of tropical crops develop rapidly. To great depth in sandy soils intensively in shallow layers in clays

38 M. A. J. Goedewaagen and J. J. Schuurman, TICSS, 4, 2, Amsterdam, 1950, 28–31

39 W. R. C. Handley, *Mull and Mor Formation in Relation to Forest Soils*, HMSO, Forestry Comm. Bull. 23, 1954

40 P. E. Müller, *Studier over Skovjord, I, Om Bøgemuld og Bøgemor på Sand og Ler.* 1878, Tidsskrift for Skovbrug, 3, 1–124 and II, *Om Muld og Mor i Egeskove og på Heder.* 7, 1884. These were later published in Berlin, 1887, as '*Studien über die Natürlichen Humusformen und deren Einwirkung auf Vegetation und Boden*

41 V. V. Ponomareva, Probl. Poch., 1962, 59–76; see also Poch. 12, 1962, 15–30, on fulvic acids in brown forest soils, neutralized by Ca and Mg. Also see M. Schnitzer et al., PSSSA, 26, 1962, 362–5

42 S. V. Zonn and D. F. Sokolov, Trudy Lab. Lesoved. 1, 1960, 61–85

43 E. Welte, ZPDB, 46, 1949, 244–78 and V. V. Ponomareva, TICSS, 6, V. 33, 1956, 207–12

5
Climate and soil formation

CLIMATE functions as an independent factor in the development of the weathering complex and of soil profiles at group level. It also causes the daily changes in the internal *soil climate*. The soil and the atmosphere interact with vegetation to form *microclimates*; the atmosphere acts on the ground to form either *surface climates*, or *site climates* if there are changes of aspect or katabatic air movement. These climates, too, are partially independent, not merely transitions from atmospheric to soil climates.

The microclimates of open surfaces are more extreme than those of either air or soil; those of woodlands are usually less extreme. Climate also indirectly influences the soil by determining the mass and form of plant production, the soil water balance over long periods, its temperature, and the rate of decay of organic matter.

Radiation. Of all climatic influences, light is the most fundamental. Many regard it as the ultimate determinant of the rate of soil formation—'for many soil processes the supply rate of incoming radiation sets the upper limit of the rates of change in the soil'.[1] Radiation is measured in gram cals/cm²/day and varies with global factors such as latitude, season, cloudiness, and local factors such as dust content and air pollution.[2] High radiation zones are the tropics rather than equatorial latitudes.[3] The tropics, including hot deserts, have high values (>600 ccd) from latitudes 10°N to 35°N in June and from 10° to 33°S in January. Absolute monthly maxima reach 750 ccd in N. Chile and central Australia in December, and in S. Iraq and Cyprus in June and July. The equator has low values, 400 ccd in most months, equal to the July maximum for Britain. Low radiation values also occur in monsoon lands—S.W. India.

coastal W. Africa and Indonesia—with monthly maxima of 550 ccd and minima of 250 ccd.

Only 1% of radiation energy is used by plants.[4] Part of the annual incoming radiation heats the atmosphere—2% in western Europe. A further 27%–31% is used in evaporation, only 2%–3% in heating the soil. Reflexion uses 20%–23%, and back radiation 34%–48%, depending on the colour of the ground and vegetative cover.[5,6] In summer in the Netherlands[7] 75%–85% of net radiation gain is used for evapotranspiration from grass not short of water, with only 15%–25% left for heating the soil and air. In Batavia the radiation totals 108,000 cal/cm²/yr;[3] at Rothamsted[6] it is 76,000 cal/cm²/yr; at Copenhagen c 88,500,[5] where the net radiation of c 30,000 cal/cm²/yr covers the energy consumption by actual evapotranspiration, leaving 2000 for loss by transport elsewhere.

The colour of the ground determines its response to radiation. Dark ground has low *albedo*, a term describing the degree of whiteness and defined as the % radiation lost to the surface. A black surface has *low* albedo (0); bog and muskeg 8; wet sand 9; dry sand 18. Deserts have 24–28; grasslands 14–37; old snow 47–50; fresh snow 83–90.[8] The lower the albedo the greater the heating of, andeva poration from, the surface.[9] Thus forests have only slightly more radiation available for evapotranspiration than grass, using more water solely through their deeper root system.[9] Further sources of energy for soil are dew and frost, important in deserts and in cold climates respectively. Only locally can inherent soil energy—the swelling of clays—rival that from the atmosphere.[10]

Soil temperatures closely follow radiation conditions, especially under clear skies. The surface of bare soils is most responsive, with a more extreme climate than the air. Cultivated soils have higher temperatures than uncultivated, and forest and grass covered soils have lower seasonal and daily ranges of temperature than bare soils.[11]

The interactions of soil and climate are complex, and it is hard to separate out the effects on soils of external climate from the independent soil climate. In many areas, soils have a greater reciprocal effect on local or on continental climates than is often appreciated.[12] For example, in humid temperate areas, 30% to 80% of annual rainfall is evaporated, in semi-arid areas from 70% to

100%. The vapour obtained is reprecipitated elsewhere, particularly in continental interiors. Even in N.W. Germany only 367 mm of a total fall of 771 mm rain is derived from oceanic air masses, the rest coming from evapotranspiration.[18] Nor is the water lost in evapotranspiration directly related to total rainfall, for it is often uniform over large areas (Table 8) and is, for example, c 400 mm in western Europe.

Soil temperatures also have a reciprocal effect on zonal and local climates. As soil is heated in the daytime it conducts and convects this heat to the air above. An extremely steep lapse rate develops in the lowermost air layers, causing turbulence, and gusts increase the removal of vapour from the soil, wind erosion on bare fields and the spread of pollen. At night the soil and lower air cool by conduction, causing gravity flow of cold air; the soil climate at one point may therefore affect that at another, for example in frost hollows or by inducing condensation as dew.[12]

The temperature of the surface soil comes to affect the subsoil, for heat penetrates downwards rapidly in hot weather, and rather more slowly in cloudy conditions or into wet soils. Normally heat penetrates at a rate of 25–75 mm/day, though percolating water may heat—or cool—soils more rapidly. Yet the generally slow transmission of heat shows why only a thin surface layer of soil heats up readily and why subsoil (1 m) temperatures vary only slightly during the year, in Britain reaching their maximum in October. It also explains the slow recession of frost from soils, which remain unmelted at depth for long periods.

Soil temperature varies widely over the globe[17]—in the humid tropics it is 25°–30°C Ø; in the dry tropics 35°C Ø; in temperate lands 18°C Ø; in the Arctic 10°C in July. Daily extreme maxima reach 40°C in Britain on bare, dark coloured mineral surfaces. On similar materials in deserts 100°C is possible. 52°C in August for a soil surface at 1300 hr, and of 34°C at 5 cm depth at 1400 hr, is reported for a black meadow soil in the Black Forest,[18] which irrigation decreased to 30°C within a short time.

Absolute minimum temperatures occur in permafrost, with −40°C Ø at Resolute, NWT (75°N 96°W).[19] Monthly means vary, with −32°C at the surface in December and −24°C in March at 1 m depth. The monthly maximum at the surface is +6°C in July and $-1\cdot8°C$ at 1 m depth. These extreme values may be compared with extreme air temperatures: $+58°C$ for Azizia, Libya; $-90°C$ near Vostok II, Antarctica $(78\frac{1}{2}°S)$ and $-78°C$ at Oimyakon, Siberia $(63°N)$.

Generally in autumn and winter in temperate areas the soil is warmer than the air; but it is then colder than the subsoil, which in turn is coldest in spring. For most of the year in Russia the soil temperature at 25 cm depth is higher than at 2 m in the air. Only in April and in July is the soil cooler than the air. The greatest difference is in Asiatic USSR, where the soil is 19°C warmer than the air in January. It is only in southernmost Russia that the soil is warmer than the air throughout the year.[20]

Aspect. In the northern hemisphere south-facing slopes heat up more markedly in summer and are also favoured in spring.[21] Yet this is a surface effect, for (in Germany)[18] a south-facing slope in late June may heat up from 41°C to 55°C between 1100 and 1400 hours and the soil at 5 cm depth only from 22°C to 24°C. An E.N.E.-facing surface cooled from 39°C to 37°C and was also warmer than the S.-facing slope at 0400 hr. The warming of southerly slopes is therefore transitory and superficial, influencing plant growth and mineralization of organic matter rather than the mineral soil. At $4\cdot2°C$ mineralization ceases; at 5°C the rate of mineralization is slow; it increases gradually to 30°C. Between 30°C and 45°C, N-production increases with an optimum at 35°C. Microbial activity decreases markedly at higher temperatures, ceasing at 80°C.

Soil freezing is an intriguing phenomenon. Frost penetrates to twice the depth in bare or ploughed land as under grass and eight times deeper than under forest.[22] Under a thick, compact litter the soil may not freeze at all, and frozen layers last longer under bare soil, peat, or grass than under forest. Snow prevents the penetration of frost into forest soils which often freeze only after the snow has melted.

Soil freezing is a function of the relation of daily maximum and minimum temperatures, not of any mean value.[23] Freeze-thaw action is also a function of moisture content, the time taken for the establishment of freezing being a linear function of increasing moisture content. Freezing is further hastened if free water can migrate from a deeper unfrozen layer to the superficially freezing layer. *Thaw* is usually more rapid than freezing, proceeding from the surface downwards if all the soil is frozen, and moving downwards and upwards if some lower layer is left unfrozen and moist.

The main effect of freezing is *soil heaving* when water expands by 9% of its volume. Yet water rarely occupies all the pore space in soil and its expansion may not be as disruptive, in either rocks or soils, as is often believed. The upward movement of water towards a freezing layer may exert even greater pressures. Organic soils show twice the heaving of sands; it is >50 mm/yr in north German winter conditions, compared with 42 mm in loess and 23 mm in sand.[24] The frequency of freeze-thaw cycles is also held to be disruptive of rocks and soil. This frequency is greatest, not in the arctic which has only 10–15 cycles per year, but in continental interiors at 45°–50°N, with 50–75 cycles per year between −2°C and +1°C.[25]

The thickness of an *ice layer in soils* is most influenced by water content, soil structure and duration and intensity of frost. Rapid freezing gives separate small grains of ice in the soil; slow freezing produces thick layers of ice near to the surface, with the subsoil ice-free, for its moisture has migrated upwards.[23] Thus with slow freezing the soil may lose its aggregation.[26]

Permafrost is a substrate feature covering a fifth of the world's land surfaces. It influences many soil groups in tundra and taiga lands, as well as in mid-continental grasslands and in mountains. Its distribution does not coincide with the extent of former ice sheets for it is best developed in poorly drained areas, with short cool summers and cold snow-free winters.[27] It can be long-established or can form on recent alluvium in the arctic. Sporadic permafrost is found as far south as Fraser, Colorado, at 3050 m (40°N).[28] It is virtually absent from oceanic areas such as Norway, though in Telemark (60°N) bedrock is permafrosted at 1500 m altitude;[29] this is termed *dry permafrost*. Permafrost is a state of ground, much thicker than the parent material (<400 m), thawing at the surface in summer to form a wet *active layer* in which pedogenic processes are usually more than counterbalanced by *cryoturbation*.[30] Permafrost need not limit plant growth for, near Verkhoyansk, summer warmth permits the growth of 20 m high conifers in the thawed layer.[31]

The active layer is usually thin, from 2 cm to 1 m in the area of continuous permafrost, to several meters in sporadic permafrost. The base of the active layer is known as the *permafrost table*. Annual refreezing of the active layer penetrates downwards, trapping its lower part and subjecting it to strong hydrostatic pressures. Involutions develop and the trapped layers may swell and burst through the frozen surface, most frequently during late winter. In summer this extruded material dries and salt incrustations form—indicating the slight chemical weathering of the parent materials. Soils of permafrosted areas are considered in chapter 12.

Wind acts in various ways upon the soil, usually destructively, by removing finer particles from unprotected soils which become coarser in texture. Organic matter content and nutrient status are decreased and living conditions for micro-organisms deteriorate. Textural analysis of wind-affected soils shows that silt is the main calibre of particle moved.[32]

Fog may influence soil formation, lowering temperatures and increasing humidities in some areas, thereby reducing evapotranspiration. On tropical mountains at 2000–2500 m, moss forest induced by fog grows on humic ferrallisols. Even the desert lithosols of western Peru are varied by winter coastal fog to produce grass-covered *lomas*.[33]

Hydrothermal zones

The pioneer Russian soil scientists showed that, over regions or zones of European Russia, with the lapse of considerable periods of time—thousands of years—climate became the dominant factor in soil formation. Certain soil profiles appeared to be characteristic of distinctive climatic regimes, which were distinguished in terms of increasing temperature and decreasing rainfall from north-west to south-east. They also observed that soils derived from the same parent material varied if produced under differing climatic conditions; while soils produced from different parent materials within the same climatic regime, given equivalent drainage conditions, eventually attained similarity in all but minor details. Parent materials, however, must not be greatly disparate, and Dokuchayev insisted that podzolic soils would be found on sands in the chernozem zone.

It was early recognized that the relation of precipitation to evaporation was critical in the development of soil profiles. The Transeau ratio P/E, an index of the efficiency of rainfall, was suggested in 1905.[34] The higher the ratio the greater the amount of water moving into the soil, dissolving and transporting substances and removing them into the drainage waters. The longer the period with no rainfall deficit (ratio >1) the more intense is the leaching. Though the ratio is not accurate at all latitudes or where rainfall occurs mainly in the warm season, it formed a basis for later and more refined estimates.

In arid and subarid areas moisture rarely penetrates deeply and subsoils are usually dry; weathering products are not removed from the soil, but redeposited at a depth appropriate to their primary solubility, and to the rainfall effectiveness. The more soluble the material the deeper it is moved—or the more it dominates the chemical composition of the drainage waters. Thus redeposition at depth at the drying front of percolating water is as important in soil profile development as the transport of materials from the upper horizons.

A grouping of soils into *hydrological series* has been presented by Voloboyev (Fig. 4),[35] who termed a modified P/E ratio a *hydrothermal range*. If the ratio is <0·2 then the soil zone (A) is *extremely arid*; if 0·2 to 0·4 (B) *arid*; if 0·4 to 0·75 *moderately dry*; if 0·75 to 1·2 *moderately moist*; if 1·2 to 1·95 *moist*; if 1·95 to 2·9 *very moist* (F); and if >2·9 the hydrorange is regarded as *especially moist* (G). The system may be placed on co-ordinates and recorded sites of soil groups related to them, showing that two or more soil groups may exist in one climatic area (Fig. 4).

In 1949 Prescott proposed a ratio $P/E^{m=0.75}$ of use for determinations of diurnal water balance or for calculations of irrigation requirement for short periods.[36] Examples of this ratio are 2·4 for rice fields and 1·2 for most crops and pastures; <0·4 represents drought.

Penman has calculated evaporation from standard meteorological observations and for various surfaces of ground by relating them to the evaporation from a free water surface, using factors of 0·6 to 0·8 to convert loss from water to loss from bare and cropped soil surfaces.[6]

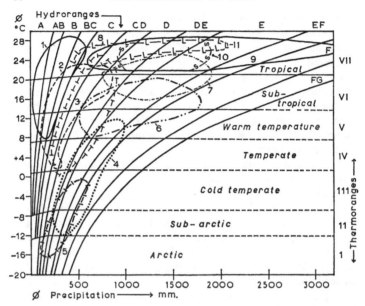

Fig. 4 Voloboyev's system of soil hydro-thermoranges. Thermoranges are named; see text for hydroranges. The overlapping boundaries of the zonal groups are: (1) serozem and aridisols; (2) chestnut; (3) (⊥) chernozem; (4) podzols; (5) tundra soils; (6) braunerde; (7) subtropical red/yellow soils; (8) fossil laterites; (9) ferrallitics; (10) ferrisols (i.e. red and brown savanna soils); (11) lateritic soils

The use of drain gauges and lysimeters also provides estimates of downward drainage losses and crop consumption. Thus *actual evapotranspiration* can be found, by subtraction, by using a water balance equation:

Precipitation=Surface runoff+vertical percolation+actual evapotranspiration+ $\triangle$ $U_{i,o}$ (gain and loss by lateral seepage)[37,38] which may be written

$$P = R_o + U_v + AE + \triangle U_{io}.$$

Though geographers are more aware of the principles of potential evapotranspiration,[39] figures of actual evapotranspiration are more relevant for non-irrigated soils. Yet there is risk of discrepancy for most figures of AE apply to water surfaces and the slightly lower figures of evapotranspiration from a bare soil or vegetated surface are far more relevant to the amount of surplus rain or potential leaching water.

Such figures (Table 8, p. 62) show that evapotranspiration does not vary greatly, either annually or regionally, like rainfall does; it has a clear-cut zonal distribution with c 500 mm from water in temperate lands; 350–400 mm from bare soil and slightly more from cropped land and 20–25% more from forest. In the subtropics AE from water is c 800 mm and 1500 mm in the tropics. Thus the amount left over for infiltration, leaching, soil storage or runoff—the *water surplus*—is most variable. There is a vast surplus in humid subtropical areas and in mountains. In humid subtropical areas the surplus is used as infiltration, in mountains as runoff; while runoff is very low in savanna and warm continental areas. Pereira,[40] in East Africa, showed that increase in the size of crop or plant does not lead to any significant change in the amount of water used (though grass transpires less than forest) and that this amount was only 90% of the evaporation from a water surface.

Evaporation from bare soil and evapotranspiration from all but deeply rooted plants are surface phenomena; evaporation is limited to the top 10 cm of soil, and evapotranspiration rarely goes below 0·7 m for all but the largest plants. A corollary is that the more frequently a soil surface is wetted by a few showers the greater the cumulative evaporation, and evaporative loss may be more a function of frequency and intensity, rather than amount, of rain.

There are three stages in evaporation from bare moist soil. The first is at a steady rate for a short time corresponding to the PE of the environment and only in a surface layer wetted in excess of field capacity. When this is achieved, evaporation decreases. The third stage has a very low rate, depending on moisture reserves at shallow depth.[41]

Evapotranspiration is a measure of the interaction of climate, plant and soil in which the role of the soil is largely passive. The portion of rainfall moving into and through the soil is a more direct factor in soil formation, promoting vertical or lateral leaching, weathering, ground water recharge and eventually stream discharge.

Rainfall will be less effective with a summer rainfall maximum, for most will then be used by plants. If there is a winter maximum, relatively more will be available for leaching (Table 8).

TABLE 8

AMOUNT OF RAINFALL AND ACTUAL EVAPOTRANSPIRATION (MM) FROM VARIOUS SURFACES IN VARIOUS ENVIRONMENTS[14]

| Location | Rain | Evaporation from | | | | Moisture surplus |
		Water	Arable	Forest	Grass	
UK[15]						
S. Pennines	1363w	600	401	c580		783P
E. Midlands	643s	556	328	397	345	87/315P
Denmark[16]						
W. Jutland	700	550	360	c450		240P
Copenhagen	550s	c550	c400	430		100/185P
W. Germany[13]						
Mean	771s	404			370	g401/180P
Harz Mts	1253wf			579		f674
„	1221wg				520	g701
N. Japan				565		f1790
Karambuchi	2355s		400b			b1955
USA N Carolina	1540s	763				777
San Dinas, Calif.	1179w	565	417b	549		b762/b706P
Manhattan, Kan.	801s	756				45
USSR Voronezh	380s		270			0—110/90P
Congo						
Savanna	850s	820			360	g490
Wooded savanna	1510s	1173				337 (+?)
Rain forest	1700s	1500		1039		f661

Notes: w=winter rainfall maximum; s=summer; f=forest;
b=bare soil; g=grass; P=percolation

The entry of water into soil and surface runoff

All this surplus of rain may not enter the soil, for additional factors control the rate of entry of water into soil. *Intensity of rainfall* is the main factor and the *entry rate* is roughly inversely proportional to it. The more intense the rain, the less the infiltration and the greater the possibility of runoff. Infiltration from cyclonic and relief rain tends to be greater than from convectional rain. The rate is also influenced by closeness of plant cover, degree of compaction of organic horizons and physical properties of the surface soil— its texture, aggregation, initial wetness and permeability. Minimum infiltration rates vary from practically nil on wet swollen clays to

2 mm/hr on clay loams, 8 mm/hr on loess and 15 mm/hr on deep dry sands.[42] These rates decrease progressively with further wetting, for soil aggregates break down and pore spaces become blocked by fine material, the surfaces becoming 'puddled'.[43] Most English lowland soils do so if rainfall exceeds 2 mm/hr.

It is probable that a deep, well drained, sandy soil can take in about 250 mm/day. Much greater falls are known from many parts of the world, though with a recurrence interval >50 years. Miami has experienced 380 mm/day, Key West 500 mm/day in a nontropical storm.[44] Zomba, Nyasaland, received 510 mm in one day in 1946. A fall of 1·03 inches (26 mm) in 1 minute has occurred in California.[45] Though burst intensities >100 mm/hr occur in many areas, it is doubtful if their erosive effect is greater than that of long continued, less intense falls, for wet soils have less resistance to erosion than dry. Many parts of the world receive falls of 125 mm/day with a return period of only 5 years,[44] which far exceeds the infiltration capacity of loams and clays.

The kinetic energy of falling rain is considerable.[46] In a 30 min storm more than 100 tons/acre may fall, striking the soil with a velocity of 20 mph, the energy exceeding 2 million ft lbs/acre for the period. Falling rain has its own erosive action, for a large raindrop striking the soil bursts upwards and outward, carrying soil particles with it. On falling, the particles gradually move down the slightest slope. When rainfall intensity exceeds infiltration, excess water moves downslope and is kept turbulent by drop impact and the soil's surface roughness; its velocity is a function of depth of water (usually a thin film), and degree of slope. This *sheet runoff* is usually channelled into linear and closely spaced micro-watercourses which erode *rills* or *denudation streaks*.[47]

The volume of water infiltrating into soil is very variable. In arid and sub-arid areas little infiltrates. Subsoil water is absent and groundwater only influences the soil in low-lying basins of a geologic or geomorphic origin. Where rainfall is high, infiltration and both continual stream discharge and dry season groundwater recharge occur, the true signs of a humid climate. The amount actually percolating per year is known for very few parts of the world (Table 8) and seems to be <50% of the total rainfall for most areas. In high rainfall areas the humus cover may lower infiltration; other factors can decrease surface penetrability, so increasing runoff or the consumptive use by plants.

The physical effect of a forest cover in cutting down infiltration is very great. Outside a spruce plantation in Lancashire (54°N),[48] rainfall at ground level was 1040 mm, at treetop height 930 mm. The 'blocking effect' of a forest is therefore 11% of rainfall. On the forest floor rainfall varied from 427 mm to 585 mm, and some 7% of the rainfall at the canopy ran down the tree boles. There is, therefore, a *throughfall* of 500 to 660 mm, and a wastage of 38% by interception and evaporation from leaves, with an additional loss of 250 mm, as transpiration, derived from the water stored in the soil. Thus only 30%–40% of the total rainfall moves into the soil in a cool temperate oceanic climate under coniferous forest,[49] though its 'leaching' effect is greatly increased by the organic acids derived from the forest debris.

A plant cover is not necessarily a protective influence to a soil though it may reduce rainfall impact. Water coursing down tree boles and dripping from large leaves may cause channelling. Soil loss may be decreased by stoniness as much as by plant cover.[50]

There are many other types of erosion resulting from the presence of water on and in soils. Isotope tracers have revealed visually imperceptible micro-solifluxion on wetting and drying of brown forest soils, the start of an erosive process.[51] Subsoil erosion— piping and tunnelling—occurs under the indurated horizons of podzols and laterites, also in alluvial material with highly dispersed subsoils. Similarly, deeper than normal penetration of rainfall into subarid soils may dissolve horizons rich in Na or $CaCO_3$ and cause collapse of the surface.[52]

The proportions of rainfall intercepted by evaporation or caused to flow as runoff may be assessed and their effects studied. Yet for most of the time rainwater infiltrates into the soil. Only once in 50 years will it briefly flow over the surface of a sandy soil; on clay soils water may flow over the surface for some short time at least once a year. However there are many natural and artificial influences decreasing the effect of this runoff. Many clays form flat or low angle slopes in low lying areas, normally well vegetated, and with absorbant mull layers. Clays may also be protected by a sheet of flood water in wet periods, or absorb water into cracks in dry periods when the incidence of convectional rain is greatest. Often clays are covered by a veneer of permeable colluvium which increases the surface receptivity, though layers on an impermeable substrate are usually readily eroded.

Snow influences soils, acting as a store of moisture and, if thawing slowly, moistens the upper layers. Rapid thaw causes erosion on steep slopes and on bare fields. Some Russian writers regard snow as a semi-permanent soil horizon; it can be ploughed or compacted to prevent rapid thaw, and it supplies not only moisture, but 1 ton/ha of nutrients and 9 tons/ha of organo-mineral particles to the soil.[53] Rainwater also acts as a source of nutrients to the soil.[54]

Thus climate acts through a synthesis of temperature and precipitation on a zonal scale. Many local effects, however, are caused by special aspects of climate.

1 D. H. Yaalon, ref. 1, chapter 1

2 H. Lundegårdh, *Klima und Boden*, Fischer, Jena, 1957, Kap. 2. Also G. A. Crabb, Michigan State Coll., Tech. Bull. 222, 1950

3 J. N. Black, Arch. Met. Geophys. u. Bioklimat. 7, 1956, 165–89

4 P. Boysen Jensen, *Production of Matter in Agricultural Plants and its Limitation*. Kgl. Dansk. Vid. Selsk., Biol. Medd. 21, 2, 1949

5 The figures are ranges from various texts. Notably, H. C. Aslyng, TICSS, 7, I, 13, 1960, 179–87; (6) H. L. Penman, Proc. Ryl. Soc. A, 193, 1948, 120–45 and NJAS, 4, 1956, 9–29

7 D. W. Scholte Ubing, Medd. Landbouwsch. Wageningen, 10, 1959, 1–93

8 C. I. Jackson, A.M.G.u.B. 10, 2, 1960, 193–9; (9) C. W. Thornthwaite and F. K. Hare, Unasylva, 9, 2, 1955, 51–9

10 B. P. Warkentin et al., PSSSA, 21, 5, 1957, 495–7 and W. W. Emerson, JSS, 14, 1, 1963, 53–63

11 R. Geiger, *The Climate Near the Ground*, Harvard UP, 1950; K. J. Kristensen, Oikos, 10, 1959, 103–20; J. R. H. Coutts, Q.J.R. Met. Soc., 81, 347, 1955, 72–9 and JSS, 14, 1, 1963, 124–33

12 H. E. Landsberg and M. L. Blanc, PSSSA, 22, 6, 1958, 491–5

13 R. Keeler, Forsch. zu Deutsch. Landeskunde, V. 57, 1951

14 Figures based on H. L. Penman, *Vegetation and Hydrology*, C.B.S. Tech. Comm. 53, 1963; from (15), G. H. Dury, *The East Midlands and the Peak*, Nelson, 1963, and from (16), H. C. Aslyng and K. J. Kristensen, Ryl. Agric. Coll. Yrbk., Copenhagen, 1958, 64–100

17 Jen-Hu Chang, AAAG, 47, 1957, 241–9

18 G. Reichelt, A.M.G.u.B., B, 6, 4, 1955, 374–99

19 F. A. Cook, Can. Geog. Bull., 12, 1958, 5–18

20 Y. S. Rubinshteyn, Poch., 10, 1960, 27–30

21 A. W. Cooper, SS, 90, 1960, 109–20

C

22 J. van Eimern, Bayerische Landwirt. Jb., 39, 1962, 1011–5

23 M. Silanpää, Can. JSS, 41, 1961, 182–7

24 W. Kreutz, Z. Acker u. Pfl.Bau., 113, 1961, 181–206

25 J. K. Fraser, Arctic, 12, 1, 1959, 40–53

26 S. Andersson, Lantmannen, 1, 1961

27 S. Taber, BGSA, 54, 10, 1943, 1433–1588. Also *Permafrost*, Tech. Mem., 49, 1957. Nat. Res. Cncl. Canada

28 J. L. Retzer, see ch. 10, ref. 19

29 J. Låg, N. Geol. Undersøk., 208, (*Geology of Norway*), 1960, 472–82

30 See references on cryopedology, ch. 12, ref. 3

31 A. A. Rode, *Soil Science*, IPST, p. 302 and 306–7; and J. S. Joffe, *Pedology*, p. 443–6

32 See ref. 12

33 M. Drosdoff et al., TICSS, 7, V. 13, 1960, 97–104

34 E. N. Transeau, American Naturalist, 468, 1905, 875–9. See B. B. Polynov, S and F, XIV, 2, 1951, 95–101, for a more modern interpretation of this effect

35 V. R. Voloboyev, TICSS, 6, V. 29, 1956, 181–88, also Sov. SS, 11, 1961, 1167–70

36 J. A. Prescott, JSS, 1, 1949, 9–19; and Int. Grassland Cong., 7, N.Z., 1956, 1–8

37 J. M. Lyshede, Folia Geog. Dan., 6, 1955

38 J. R. Mather, Bull. Am. Met. Soc., 35, 2, 1954, 63–8

39 J. R. Mather, *Measurements of Potential Evapotranspiration*. John Hopkins Univ. Pub. in Climatology, 7, 1, 1954

40 H. C. Pereira and P. H. Hosegood, JSS, 13, 2, 1962, 299–313

41 A. Feodoroff and M. Raf, CRAS, Paris, 255, 1962, 3220–2

42 T. J. Marshall, *Relations between Water and Soil*, CBS, Tech. Comm., 50, 1959

43 W. W. Emerson and G. M. F. Grundy, J.Ag. Sci.,44, 3,1954,249–53

44 D. M. Hershfield and W. T. Wilson, J. Geophys. Res., 65, 3, 1960, 959–82

45 Quoted from ref. 44

46 H. Wischmeier and D. D. Smith, TAGU, 39, 2, 1958, 285–91, and Rainfall and Erosion, *Adv. Agron.* 14, 1962, 109–48

47 M. Donzwalski, Acta Geog. Univ. Lodz, 8, 1959, 67–97

48 F. Law, J. Brit. Waterworks Assn., 38, 1956

49 J. D. Ovington, Forestry, 27, 1954, 41–53

50 G. Rougerie, CRAS, 246, 1958, 290–2

51 B. Kazó and L. Gruber, Agrokem. Talajt, 9, 1960, 517–26 (abstract in S and F, 1471, 24, 1961, 216–7)

52 On many features of erosion see F. Fournier, *Climat et Érosion*, Pr. Univ. Fr., 1960

53 I. N. Stepanov, Poch., 3, 1962, 44–52

54 H. Riehm, Agrochimica, 1962, 2, 174–88 and E. Eriksson, Tellus, 4, 1952, 215–80

6

Geomorphic factors in soil development

THE classical exposition of soil-forming factors referred to 'relief' and regarded it as static, and Jenny[1] regretted that more was known of the effect of 'topography' in the erosion of soils than in their formation. Contrasting soil formation on dry knolls and in moist depressions, he implied that soil climate is largely related to the water table, which in turn is dependent on relief. He considered truncated soils on slopes and buried soils at the foot of slopes as a 'toposequence'. Quoting Marbut, he referred to the concept of 'normal' soil development on undulating topography, and to 'not normal' soil development on level topography—for some horizons are over-developed (eg claypans); other soils, on steep slopes, are 'abnormal' for they *lack* some of the horizons contained in normal soils. Jenny's account of the relief factor is distinguished by its brevity.

Geomorphologists are much encouraged by Cline:[2] 'We have altered our model of soil in three major aspects during the past 25 years. These involve concepts of geomorphology, of time as a factor in soil genesis and of processes of soil formation'. Cline considers that Marbut's concept 'of normal soil seems to follow from Davisian geomorphology, with a normal relief on which there is an equilibrium of soil development and erosion such that the normal soil sinks into the parent material as rapidly as material is removed from the surface. This is consistent with the down-wearing of land forms and especially of interfluves'. It is known as *slope decline*.

Cline states that soil scientists 'have recently become more aware' of Penckian concepts; 'that the marginal slopes of an eroding upland retreat "parallel to themselves" with new and lower erosion surfaces developed on which sediment may be deposited, the old land surface remaining relatively stable. This has startling implications for our model, for soil does not reach equilibrium with environment, but changes to an ever older product until *slope retreat* consumes it'. Thus soils on recent slopes may vary greatly but, 'on the oldest unconsumed slopes, the soils never achieve an equilibrium with the formative factors for the rate of change of soil becomes extremely slow and the soil is often related to past process influences rather than to those currently operating'.

Recent studies have, however, shown that Penck did not postulate parallel retreat of slopes,[3] which term is more appropriate to the theories of L. C. King.[4] Penck's views are best categorized as '*slope replacement*'. We now know that Penck regarded 'the development of valley slopes as the fundamental natural measure of erosion during the incision of a stream . . . if a river's rate of downward erosion increases, valley-side slopes become steeper in successive units of time and convex slopes result. As erosion intensity weakens, the reverse is found and concave profiles result'.[3]

Differences in soils are often related to the combination of slope facets into convex or concave elements. Much of the geomorphic mystique attached to these terms disappears on regarding a concave slope as a steep slope surmounting a low-angle slope, the inverse for a convex slope. Soils in the lower parts of concavities, where moisture and fine colluvium accumulate, have deeper profiles than the upper steeper parts. Yet deepening on lower sites does not proceed indefinitely for (1) the supply of material from the upper slope diminishes on its retreat and (2) the fine superficial material on the lower site is readily erodible. On convex slopes erosion is most intense on the lower steeper slopes, and is relatively active on upper gentler slopes when compared to the lower parts of concave slopes of equal angle. Convex slopes are often eroded to bedrock, the lower facet rapidly receding by physical weathering to consume the upper, lower-angled facet, creating a basal concavity in so doing.

In the early days to which Cline refers, soil scientists agreed that relief or 'topography' was a modifying rather than formative influence on soil, affecting erosion and deposition of material and soil moisture regime. Thus slope and drainage conditions were the most studied aspects. Just as Davis pictured landforms developing to a final end product and 'knew in his own mind, from the start, the type of landscape his deductions ought to produce',[5] so the extreme view of soil development was that once all the factors were known one could deduce the final profile—'Such a method, whatever its merits, is hardly justified in natural science'.[5]

Bakker has defined geomorphology as 'the science which investigates the landscape forms, the weathering processes, the erosion of weathering products and the sedimentation which are closely bound up with the comparative climatological or paleoclimatological context of these phenomena'.[6] Thus the emphasis of geomorphology changes from qualitative description of landforms—and inference as to their sequential development—to quantitative analysis of slope processes; exact determination of the present balance of weathering and of erosion; and the sequence of erosion and deposition in past times as proved by stratigraphy. As there is a close connection between the soil and the age, relative position and equilibrium of all slopes in the natural landscape it is desirable to integrate such studies, for anyone wishing to understand the soil must also know the type and rate of slope and landform development.

A simple and elegant approach to slope equilibrium is made by Jahn.[7] He divides slope processes into two groups: (1) the climatically controlled alteration and crumbling of rocks and (2) denudation. The first decreases the rock by solution, pore space is created and hydration takes place. The weathering mantle formed deepens progressively unless the denuding forces of gravity, water, wind or ice remove it. These forces achieve minimum corrosion of rock, transporting fine material parallel to the surface, either leaving behind coarse material which is altered by weathering only with difficulty, or else re-exposing the rock, when weathering recommences. Thus one may conceive of a *balance of denudation* on a slope which has a value—the relative depth of the soil—which depends not only on the potential decomposition and transport at each point, but also on the relative position of that point on the total slope, for the portion upwards from it is a potential source of material.

If the rate of soil formation be termed 'D'; 'S' the surficial removal by active agents of denudation; 'M' the mass movement by gravity; and 'A' the material added from upslope, then, where

D=S+M+A, the thickness of soil remains constant and one has a *balance of equilibrium*. If D is low (D<S+M+A) there is an *active balance* of denudation which state can only exist until bare rock is exposed. In the third case (D>M+S+A) soil depth increases by rapid weathering or by addition of colluvium. This is termed *passive balance*.[7] With equilibrium balance the slope is clearly adjusted to maintain a steady state between the resistivity of the material to erosion and the intensity of erosional processes. The state of balance changes most rapidly with time if the parent material is unconsolidated or friable, and only slowly if compact. On slopes underlain by compact materials active balance is present on the upper parts of slopes, passive at their base, as both water—promoting vertical corrosion—and denudation products accumulate there. On friable materials lower slopes have active balance, the upper parts a passive balance.

The concept of denudation balance is a descriptive one but is capable of quantitative analysis and dating. Any landform thus comprises contrasted slope areas or 'facets' each with its characteristic slope angle, partially a response to the type of parent material, partly due to the denudation processes acting on it. Each facet contains several segments each with its own state of balance represented by the depth of soil. Soil depth is never uniform over one simple slope facet and so one cannot infer soil process from slope form, though one may safely infer soil process from the balance on any one part of a slope as well as the relative thicknesses of A, B and C horizons.

Slope evolution (active balance) in most climates occurs in limited upland areas in response to intense removal of material. Most areas in lowlands have passive balance due either to rapid corrosion or to intense alluviation. In the tropics passive balance obtains on old unconsumed interfluves. In deserts and periglacial areas there is active balance on many slopes, and localized accumulation in depressions, but the balance in deserts is also a matter of position in relation to larger regions of wind erosion or redeposition. Slopes of similar angle and lithology may therefore show different states of denudational balance on different landforms or under contrasted climatic conditions. Hence climatic geomorphology and soil regionalization come together on both local and zonal scales.[8,9]

Slope replacement and soil formation. A valley side slope, when first formed, is convex in profile, with active balance at the foot and quasi-equilibrium in its upper part. As valley deepening wanes the floor is widened, the side slope changing to a concave form, with passive balance at its base and active balance in its upper parts. Side slopes diminish in relative height and are eventually replaced by the upward growth of a valley side pediment, defined[10] as 'an erosion surface lying at the foot of a receded slope, underlain by rocks or sediments of the upland, either barren or mantled by pedisediment and with a concave upward profile longitudinally'. In time soil formation begins on pediments. The higher parts are younger surfaces than the lower, less-recently cut, parts. Yet the lower parts may be covered by alluvial material, the centre by wash material and the upper parts by colluvial material; they can therefore have younger parent materials than the actual age of the erosion surface would imply.[11]

From this analysis one may formulate a *relief catena* (Fig. 14, 5) with the elements: (1) the upland or *crestal slope*; (2) the pediment backslope (*scarp* or free face), which is the receding slope; (3) the pediment footslope (*pediment, sensu stricto*) and (4) the lowest element, the *alluvial toeslope*.[10] Terraces may diversify the toeslope which may be further subdivided into wash, alluvial and bottomland slopes. The pediment backslope rarely completely coincides with the hard rock band forming the upland, often crossing rocks of varying hardness without change of gradient. Varying lithosols form on such slopes. Soils on the crest, near the freeface, are usually highly leached and eroded. Inward from these marginal sites, on plateaux, developed zonal soils occur, of considerable age, and hydromorphic or planosolic variants may exist in the central, poorly drained parts of the widest crestal plateaux. Zonal soils in a much earlier stage of development occur in the central stable parts of pediments, while regosols form in the upper part of pediments and on alluvial toeslopes. The oldest soils survive on the crestal plateaux for a considerable time, for the rate of retreat of major rock scarps is very slow. King cites 1 ft in 150–300 years in Natal (30°S).[4] As a geologic average for north Sweden (68°N), Rapp[12] gives a minimum of 0·04 and a maximum of 0·15 mm/yr under present conditions. Thus if we imagine a small interfluve, 5 miles wide, 2 million years must elapse before the centre is consumed in the subtropics. On geological grounds this is probably a very low estimate. When the interfluve is completely destroyed by slope retreat, the pediment segments join to form a low angle pediplain, with the youngest debris at the centre.

Slope angles and soil formation. Natural slopes with a continuous soil cover seem to be at all angles below 40°; *slope angles*, however, are by no means random in nature and certain characteristic angles have been noted.[13,14] Investigations in various environments show that 45°, 40°, 37°, 31°–32°, 26°–27°, 19°–20°, 13°, 10°, 7°, 5° and 2° are the slope angles most commonly present, as well as 'flats' of 1½° to 0°. Cliffs are considered as >50°.

The proportion of the world's land surface with angles >40° is low. Slopes >30° are rare in central Europe and the same may be said for all but the mountain areas of the world. At angles above 27°–30° fossil soils have been destroyed by denudation though largely preserved on slopes of lower angle.[15] This *angle of 30°–32° is critical for stability* in many areas, and is common on many resistant rocks.[16] Usually rectilinear, slopes from 32° to 37° intersect sharply with other facets of this group—above, below and laterally —due to thinness of soil cover. They form the limiting angle for coherent soil cover—usually a litholic AC profile—and are undisturbed by mass movement unless wetted excessively. Slopes of 37°–40° sharply intersect with each other[18] and have lithosolic (A)C soils. Slopes at 26°–32° are usually stable for long periods and have thicker AC soils, with the C horizon dominant. These slopes are not rectilinear, but slightly convex or concave—the steeper (30°–32°) segments having thinner soil. Slopes at 25°–32° pass into each other, laterally and longitudinally, without perceptible 'breaks' of slope.

However, soil stability may not exist at much lower angles (eg 7°–10°), for, as coarse debris is comminuted, increasing fines and moisture decrease the shear strength of the soil and give greater chance of *slumping* and *creep*.[17] Such slopes also show outcrop curvature and both A and C horizons are usually disturbed. While the average gradient for earthslides, mudflows, talus creep and debris avalanches is 30°, *solifluction*, a much slower process, occurs on slopes of *c* 15°, preventing the development of A horizons in N.W. Sweden.[12] *Talus creep*[18] reaches 10 cm/yr on unstable slopes, solifluction 0·6 to 5·0 cm/yr for 6° and 16° slopes respectively.[14] Obviously these periglacial and moist oceanic environments are active, but not necessarily much less active than tropical environments.[19]

Slope wash is less effective in denudation than creep, removing 0·08 cm³/cm/yr, against 1 cm³ by creep, at the same angle of slope (26°)[12]. It is ineffective at angles <20° in the presence of a plant or litter cover in a temperate oceanic environment.[17] Short slopes of 5°–7° are characteristic of the degraded margins of crests; they are also typical of many of the upper parts of pediments. Extensive rectilinear or smoothly curved slopes of such angles are rare,[13] and are nearly always at 2°–3° on toeslopes. Level land is as rare as steep slopes, confined to terraces, and to deep unconsolidated deposits of recent origin such as loess, alluvium or till which completely bury underlying relief. When such material is redissected a series of terraces and meso-pediments forms.[11]

Slopes and soil moisture. It is commonly accepted that soil moisture increases from the top towards the foot of any slope. Four influences help establish this moisture gradient: increasing depth of soil available for wetting by surface and subsurface flow; increased moisture retention capacity of soil on lower sites; decreasing evaporation on footslopes because of lowered exposure to wind; and incomplete wetting of upslope soils because of thinness and stoniness.[20]

Soils are therefore deeper at the foot of slopes because of increased subsurface weathering and the addition of material from upslope. A horizon depth is greatest near the base of slopes, compared with the upslope parts; but the *ratio of A: B+C horizons* is usually less at the foot than in the central parts of slopes. However, such conclusions apply to total slopes, not to their individual facets, and are based on latitudinal sections or 'slope profiles' rather than on segments or 'slope areas'.

It is often thought that the underlying rocks determine the area, form and angle of slopes. On a microscale and as far as soil sites are concerned this is not so. As Chorley[21] remarks, bedrock is not to be considered a parent of the related morphology, but rather a grandparent. Intervening are the external soil-forming influences, and we know that different angled slopes can exist on the same parent material and similar angled slopes can traverse two adjacent materials of different resistance.

Angle and length of slope mainly influence the relative proportions of infiltration and runoff, and the balance of denudation and soil formation. Horton's[22] theory of surface runoff is based on two relevant concepts: (1) there is a limit to the amount of water a soil can absorb in a given condition from rain as it falls—the infiltration capacity—which, when exceeded, gives rise to runoff; (2) there is a minimum length (x_c) of overland flow required to produce sufficient volume of water to initiate erosion. X_c depends on slope angle, intensity of runoff, infiltration capacity, soil resistance and other factors. Erosion rate is governed by either the transporting power of overland flow or the amount of erosion, whichever is smaller.

Thus if the volume of runoff reaches erosional strength over a length x_c, the soil would be disrupted. Yet the erodibility does not increase progressively downslope, nor do soils deepen uniformly downslope, despite many *a priori* deductions. On low angled watersheds differential accumulation of soil moisture occurs as '*percolines*'[23] producing long lines of deeper soil trending downslope from near the crest towards stream sources. These lines are paralleled and separated by zones of shallower drier soils of uniform depth downslope. Only in the percolines do soils tend to deepen downslope. Lateral subsurface flow and surface wash moves from the drier areas towards the percolines. These become saturated at times of high rainfall, or in wet seasons, and concentrated surface flow and erosion commences on their lower parts to form rills and gullies. This erosion rapidly extends headwards in the deep soils of the percolines well within the limits of the critical distance postulated by Horton.[22] The '*belt of no erosion*' is therefore not of uniform width, and the original slope surface comes to exist in a fragmented state on which highly leached but shallow soils develop; while deeper soils develop in percolines until gully erosion either truncates or totally removes them.

Soil profiles in drainage basins have been interpreted according to the *catena concept*,[24] which implies that on lowest slopes the soils assume a hydromorphic intrazonal form or a moist variant of the zonal type, and that, with increasing altitude, zonal or azonal soils result. A catena is usually defined as 'a sequence of soil profiles which appear in regular succession on morphologic features (or in similar drainage basins) of uniform lithology'. Basic to the concept is the change along relief-profiles from well drained zonal or auto-morphic soils on the crestal slope (which, if wide, will also have planomorphic soils in the centre) to excessively drained, thin, (xeromorphic) or eroded soils on the free face; to hydromorphic soils on moist sites on the alluvial toeslope and to ombromorphic or peaty soils on bottomland sites. Milne's original statement refers to 'a grouping of soils which, while they fall wide apart in a natural system of classification are yet linked in their occurrence by con-dition of topography'. He referred to a catena as a 'unit of mapping convenience'. Later he wrote 'it has become apparent that we have to deal with two classes of them. In one the parent material does not vary . . . in the other . . . the topography has been carved out of two superposed formations'.[25] (See Fig. 14, 5.)

The catena concept is a thorny one, much abused since its for-mulation in 1935, not least by geographers. Clarke[26] objected to the use of catena in England where different geological beds occur in rapid succession down the slope (eg in the North Downs) and provide a mixed catena, for at least three kinds of soil profile are found on each bed in the lower, central and upslope parts of each outcrop. Milne regarded with dismay 'the topographic–denudational sequences in which parent material is not identical throughout, but is complicated by different geological horizons having been exposed by dissection or faulting. I do not think we can (here) employ the word catena unmodified without spoiling its simplicity of connota-tion. We need another word for that kind of soil complex of some importance in the Rift Valley. . . . The concept is that of a hybrid or "misbegotten" catena, of mixed genetical origin in respect of the lithologic factor, like the last line of the famous Mendelian limerick: "One black, and one white, and two khaki"!'[26]

That soil profiles may be related to the morphology and moisture regime of the side slopes of drainage basins is evident; that they are related to the characteristics of the whole basin is also feasible. Large drainage basins of uniform lithology have longer stream segments and gentler gradients than smaller basins; are less nearly circular and have lower relief ratios.[27,28] Thus soils are deeper and more uniformly developed than in smaller basins. With uniform lithology small drainage basins, with their steeper slopes, have active balance; poorly developed lithosols dominate with limited alluvial and colluvial deposits and associated regosols on toeslopes. As lithology changes so does the morphology of the drainage basin and the proportion of developed soils. Increase in resistance of a bed results in stream drainage basins of larger area, increase in the length and gradient of streams, decrease of stream density and of relief ratio. This increases the sites available for developed zonal soils and gives less alluvium and more colluvium. Quantitative geomorphology thus permits objective comparisons of form elements of watersheds, drainage basins and sites; the actual soils developed then vary only with lithology, rainfall and aspect. Hence local variations of soil may be related to degree of slope, water balance of site and relative position in the drainage basin.

In extremely large drainage basins of sub-continental size, zonal geomorphic history and climate may have greater effect, changing both laterally and with increase of altitude. Mountain or vertical zonality is also related to latitude and to continentality—position relative to the oceans.[29] Four comparative examples of mountain zonality are given which are self-explanatory (Fig. 6, p. 122). Some tropical mountains show a regular vertical zonation, varying in hydromorphism only in response to variations of rainfall and of steepness of slope (Kivu); others show marked contrast of arid and humid slopes (Andes). Mountains in temperate and subarctic latitudes show greater contrasts on opposed slopes; of *ubac* and *adret*, moisture and dryness; and often anthropogenic activity is more intense on one slope than another. But in the less intensively eroded lowlands, slopes are more stable and develop more gradually. Such matters are best considered under 'age of landform' and are more related to geomorphic evolution than to geomorphic process. This links the geomorphic factor with that of time in soil formation.

1 H. Jenny, *Factors of Soil Formation*, 1941, p. 89
2 M. G. Cline, PSSSA, 25, 6, 1961, 442–6
3 M. Simons, TIBG, 31, 2, 1962, 1–14
4 L. C. King, BGSA, 64, 1953, 721–52
5 J. P. Bakker and A. N. Strahler, Rept. Comm. IGU, 1956, 1–12
6 J. P. Bakker, Rev. Géom. Dyn., 1959, 67–84
7 A. Jahn, Czasopismo Geog., 25, 2, 1954, 38–57 (fr. 57–64)

8 J. T. Hack, AJS, 258A, 1960, 80–97

9 R. W. Jessup, JSS, 12, 2, 1961, 199–213

10 R. V. Ruhe, Elements of Soil Landscapes, TICSS, 7, V. 23, 1960, 165–70, and (11) SS, 82, 6, 1956, 441–55

12 A. Rapp, Geog. Ann., 42, 1960, 2–3, 67–200

13 R. A. G. Savigear, TIBG, 31, 2, 1962, 23–42

14 A. Young, Zeit. Geom., 5, 1961, 126–31

15 H. Poser, Der Naturwissenschaften, 34, 1947, 10–18

16 R. A. G. Savigear, IGU Rept., 1956

17 A. Young, Nature, 188, 1960, 120–2

18 E. J. Parizek et al., J. Geol., 65, 6, 1957, 653–6

19 R. A. G. Savigear, Zeit. Geom., Supp. 1, 1960, 156–71

20 B. D. Van't Woudt, TAGU, 37, 5, 1956, 588–92

21 R. G. Chorley, AJS, 257, 1959, 503–15

22 R. E. Horton, BGSA, 56, 1, 1945, 275–370

23 B. T. Bunting, AJS, 259, 1961, 503–18 and GJ, 130, 1, 1964, 506–12

24 G. Milne, *A Provisional Soil Map of East Africa*, Amani Memoirs, 28, 1936. (Other writings of G. Milne are in: J. Ecol. 35, 1947, 192–265; Geography, 29, 1944, 107–13.)

25 T. M. Bushnell, The Catena Caldron, PSSSA, 10, 1945, 335–40 and 7, 1942, 466–76

26 G. R. Clarke, *Study of the Soil in the Field*, 4th ed. OUP, 1957

27 M. E. Morisawa, BGSA, 73, 1962, 1025–46

28 J. Tricart, Rev. Géom. Dyn. 1959, 2–15

29 V. V. Dokuchayev, *The Theory of Natural Zones, Horizontal and Vertical Zones*, 1889, in Collected Works, VI, Moscow, 1951

7

The time factor in soil formation

> Soils are not only very variable in space but also
> comparatively unstable in time. Actually we do not
> know of any soil which has retained its character
> for ever.   V. V. DOKUCHAYEV

THE other soil-forming factors are synthesized in the time factor,
the role of which may be assessed in four different ways: (1) by the
relative stage in soil development from a time zero; (2) by the rate
of formation of a unit depth of soil or soil horizon; (3) by reference
to the age of the slope, landform or weathering complex on which
the soil is developed; and (4) by absolute dating of a part of the
soil profile. These approaches may of course be combined. A fifth
approach is by experiment on fresh rock or on soil columns in the
laboratory, or by inspection of the decay of such man-made features
as building stones. One must carefully distinguish, however, between
rates of rock weathering and of true soil formation. Such experiments show that sandstone breaks down more quickly by physical
weathering than limestone does, though the reverse applies in
chemical weathering. As more material is lost by solution from
limestone, and there is less residue, the rate of formation of a unit
depth of soil is absolutely slower, even though rock weathering is
faster.

Stage of soil development will be illustrated in detail later, but
generally there are initial and early stages, with (A)C and AC profiles respectively, when soil properties are closely related to parent
material and to the activity of invading organisms. As external
factors, especially climate, increase their influence, developed ABC
profiles result. After considerable periods of time, or else in a short
time with intensive weathering and leaching, a soil reaches a stage

in which further development is very slow. According to the classical view of Marbut-Davis, a climax soil is formed, which is time-dependent, the end of a sequence of development. However, this climax form is not a short term feature and persists longer than the conditions which have brought it about, to become a relic form; or else, more commonly, retrogressive conditions occur, such as erosion. The soil, therefore, is an open system.[1]

Only limited data concerning *the time taken to form a unit depth of soil* are available. On resistant rocks this may take many centuries; on unconsolidated materials—loess and tephra—the rate is more rapid. Highly permeable materials such as outwash sands, with low contents of soluble materials, may show deep but only faintly differentiated eluvial and illuvial horizons. Bridges[2] showed that the initial stages of soil development in temperate areas were favoured by the microclimate of south-facing slopes. These had greater rates of organic debris decay and, drying out to depth, allowed deeper percolation of leaching waters. After 40 years LFH layers, weak structural forms and decalcification were evident to 40 cm. On north-facing slopes, soil processes had penetrated to only 2 cm depth.

The classic example for rapid soil formation on tephra is the 35 cm of soil formed in 45 years on Lang Eiland, Krakatoa (6°S).[3] Weathering of a fan of pyroclastic material on St Vincent (13°N) produced 1·8 m of clayey B horizon in 4000 years; the soil resembling the yellow-brown volcanic ash soils of Japan. Thus the soil had apparently formed at a rate of 0·45 m/1000 years. Similarly, in El Salvador (14°N), 1 m of ash had weathered in 2993 ± 360 years;[5] yet on Mt Shasta (California, 42°N), volcanic mudflows aged 205, 566 and 1200 years showed no alteration. Hence one cannot compile yearly or even century long average rates of alteration to soil of this or any other parent material. After deposition, the decomposition of tephra may at first be very low or even nil for a considerable time. The rate then increases while the easily soluble materials are being removed, declining as resistant minerals are encountered in increasing proportion within the weathering complex.

In the young tropical soil from St Vincent intensity of mineral grain decomposition was not so far advanced as in many podzolic soils of cool temperate regions which, with high acidity and strong leaching, alter more intensely in thin profiles, in contrast to the tropics where less intense alteration of individual grains takes place to much greater average depths.[4] Hence the total volume of mineral matter destroyed is much greater in the tropical soil per unit of time. One must conclude that apparent depth of weathering and soil horizon differentiation are not sure guides to relative amount of weathering, for intensity of weathering is as significant as its depth of penetration.

Assessments of the rate of soil development are often achieved by study of soil and plant succession on sites recently emerged from an ice cover.[6] In the Tirol (47°N) on 45 and 85-year-old stadial moraines, increase of soil acidity, of soil N and more clay formation were noted on the older material.[7] In the cold, wet conditions of Glacier Bay, Alaska[8] (59°N), leaching of soluble material and of exchangeable metal cations modified the surface and was intensified by plant colonization. In the short growing season organic matter is produced, yet does not decay; it accumulates, becoming wetter and more acid as time passes. Twelve stages have been outlined in the succession from bare rock to muskeg,[9] and at all stages accumulation of organic litter and vegetational sequence control soil development. After an initially slow start, soil formation accelerates until 45 cm of surface mineral change can be discerned on 50-year-old material. This soon ceases as far as deepening is concerned. Further chemical change, in the form of complete decalcification, penetrates to 15 cm in the next 100 years and then takes 1800 years to reach the base of the soil at 50 cm under a deep organic cover.

In many arctic and muskeg soils organic matter is mainly fibrous and undecomposed, and natural decomposition of peaty layers is very slow. The great age of organic matter in unglaciated arctic regions is proved by Tedrow[10] who found that in the non-humic part of an arctic brown soil at Point Barrow (71°N), the age of the organic matter was 2900 ± 130 years, and in the humic surface portion 2000 ± 150 years. Much lower figures obtain for organic material in temperate soils. Tamm[11] found that on land in mid-Sweden the A layers of a podzol had organic matter less than 100 years old, while the Bh horizon had humus 370 ± 100 years old, even though the site had been forested for 9000 years.

Dunes provide a means of assessing soil formation from time

zero. Salisbury[12] found the ages of ridges at Southport ($53\frac{1}{2}°N$) from old maps (1610 and 1736), and by tree ring analysis. In the first century of development H-ion concentration remained low, though $CaCO_3$ content declined from 6·3% to 1·1%, and, after 280 years, $CaCO_3$ reached zero with pH 5·5.

The tills of central Europe and the USA also provide a means for dating the rate of soil development on *surfaces of known age*. Stremme[13] reports soil formation on till of the Drenthe stage of the Saale glaciation (112,000 years BP) and on Weichsel drift (72,000 BP). The Drenthe has soils with thick, light grey, leached A horizons with Fe content of <0·59% and a yellow-grey B horizon (ie Fahlerde); the Weichsel has brown soils with uniform profiles and an Fe content in the A horizon of 0·9% (ie Braunerde).

Similar contrasts are possible of the iron-humus podzols of western Jutland (56°N, 9°E) on Saale till; and the base-rich brown soils of eastern Sjaelland (55°N, 12°E) developed on tills of the Langeland stage (14,500 BP). Yet there are dangers in such comparisons based on age of parent material and assumed available time for weathering, for one cannot say with certainty that both Danish soils are explicable solely by reference to the time factor. The parent materials are greatly different; though both are tills, one, in the west, is derived from a source material of Miocene sands while the eastern boulder clay consists largely of ground-down Senonian chalk. There is, too, an appreciable difference in present climates between the two areas, with a far greater surplus of rain for infiltration in the west (Table 8). Obviously the soil in western Jutland started as a sandy base-poor brown earth, but we do not know if its development to its present state has been progressive, or whether soils developed previous to the last glaciation were completely removed by periglacial action during the last glaciation, with an ice front only 50 km to the east. Thus the 'time zero' of the present podzols may be 16,000 BP, beginning with sol brun acide.

Apparently over the heathlands of the North Sea lowlands podzolization is a very recent phenomenon dating perhaps from early Neolithic times, an alteration of podzolic sandy brown earths related to anthropogenic rather than to climatic or regolith factors.[14] Despite the intense surface podzolization, lime-rich (10–15% $CaCO_3$) marly drift can be seen at shallow depth (3 m) in west Jutland.

In tills, lime present originally may be rapidly leached to considerable depth, causing a post-glacial soil formation of a degenerative character.[15] Apparently soil processes were not under the direct influence of lime in the last 4000–7500 years in southern Sweden. Clearly decalcification is an important process, but more intense weathering must occur before a soil can approach the podzolic state and far more before complete podzolization is achieved—a state so often used to illustrate the rate of soil formation.

Podzolization may be rapidly achieved after deforestation in humid areas, or after reafforestation by conifers upon *decalcified* sands. On psammitic brown earths in eastern Skåne (56°N, 14°E) 50–60-year-old conifers produced 4–10 cm thick podzolic A horizons (=1 cm in 6 to 15 years). Tamm cites rates of podzolization:[16] (1) on a 120-year-old sand covering peat bog at Malingsbo, Dalarna (60°N), with a thin bleached earth 1·3 cm thick; and (2) the Ragunda Lake, E. Jämtland (63°N) drained naturally but catastrophically in 1796. Studied in 1912, when, after 116 years had elapsed, 'scarcely visible podzol profiles had developed . . . chemical analysis confirmed that the process has not yet reached any result worth mentioning'. In this area decalcification occurred, to 25 cm under pine, to 64 cm under mixed forest with moss, in the same period. For another locality Tamm estimated that 'a normal podzol with 10 cm mor, 10 cm bleached sand and 25–50 cm B horizon required 1000–1500 years for development', but that older soils, from 3–7 thousand years old, did not show greater development.

For tropical areas with constant temperature, rainfall is the significant variable influencing the rate of soil development. With 2000 mm/yr it takes 22–77 thousand years to weather 1 m of calcic granite to form a 'latosol'; with 1500 mm from 53 to 102 thousand years.[17] Yet it may take only 35 years to change a lateritic clay to a brick hard laterite on deforestation.[18]

A report on soils developed on levées of the Rhine and Meuse (52°N), dating from 100 BC to 8000 BC,[19] shows that various rates of formation exist during the life of a soil. The soils do not differ greatly, for all had lost lime and showed poor horizon development. However, soils on older deposits, dating from 10,000 BC, show developed grey-brown podzolic profiles with a B_t horizon due to mechanical illuviation of clay. This profile was differentiated during the period 10,000–8900 BC, whereas during the younger Dryas period (8900–8300 BC)—the last cold tundra phase of the Pleistocene —biological intermingling of soil and humus was confined to the unfrozen surface soil and could not act more deeply. Thus a few hundred years of great intensity of factors working to depth may achieve far more in horizon differentiation than some thousands of years of less intensity—as in the post-8000 BC soils. By extension, it is clear that a few centuries in which the whole complex of environmental influences is changing is far more effective in soil development than long periods during which the factors do not change but create conditions of equilibrium.

A concept introduced by Australian workers relating to the time factor is that of the 'k-cycle', or series of 'k-cycles'.[20] Each cycle commences with an unstable phase, 'k_u', of erosion and deposition, and is concluded by a stable phase, 'k_s', in which soil formation begins on newly exposed erosion pavements and on adjacent fresh deposits. Successions of buried soils provide datable evidence and prove recurrent cycles of stable and unstable phases. In New South Wales,[21] on the coastal plain (34°S), k_u phases occur in dry climatic periods—when storm rainfall gives hill wash, gully erosion and soil creep. Soil formation occurs in humid k_s phases. Some steeply graded drainage basins and gently sloping floodplains show three cycles; k_3 has upper surfaces with red and yellow podzolic soils which began to form in the Würm I–II interstadial, 29,000 BP; a k_2 cycle has grey-brown soils initiated 3740 BP; a k_1 cycle started in 390 BP and has 'minimal prairie' soils on gently sloping flood-plains. An epicyclic surface, k_0, starting 1–120 years ago, proves the incipient destruction of the lowest k_1 surface. Thus soil development can be dated in relation to erosional history and this in turn to climatic periods.

Interesting for the geomorphologist are *buried soils*.[22] These range in age and thickness from developed soils buried under thin layers of transported material, to fragments of older inceptisols buried under later materials which have fully developed contemporary profiles upon them. Often, if organic materials remain, pollen or C−14 dating is possible. As an example, two superimposed podzol profiles were found in windblown sands in Emsland (53°N).[23] The upper podzol dated from 2500 BP; the older podzol had a surface organic horizon dated at 7000 years BP. Fränzle[24] studied complex interstadial soils on Würm loess in N. Italy. In a poorly drained area two Würm loess deposits overlying Riss moraine, itself gleyed, had developed extreme pseudogley profiles, one in an interstadial period, the other since deglaciation. The only difference was that the earlier loess soil had acquired frost-wedge phenomena during the late Würm cold period.

Relic soils are those dating from past climatic, geomorphic or geologic periods found far outside their present climatic range. They may be buried or at the surface. Examples from Ireland[25] show reddish soils, related to Mediterranean red soils in morphology and chemical characteristics, and preserved in solution hollows in the limestone terrain of south Wexford ($52\frac{1}{2}°N$).[26] Presumably formed under warmer climatic conditions of late pre-Pleistocene times, they are covered by glacial drift on which the totally different present surface soil is developed. It is also possible to find the inverse situation of grey-brown podzolic soils formed under colder conditions in presently warmer environments, for example in southern Portugal, formed during the glacial periods.[27]

There is thus no complete or simple calibration of soil formation and time. Usually the rate of soil formation varies with temperature, permeability, rainfall effectiveness, H-ion activity and biologic mixing; it varies also through time. If one considers the age of landforms and slopes in relation to the soils developed on them one is not necessarily on easier ground. All one can then provide is a maximum age of the soil and try to find evidence for any disruption within its profile.

Thinking on a continental scale one feels drawn to the synthesis made by Carter and Pendleton[28] of the soils of the humid areas of eastern USA. They categorize soils as juvenile, mature and senile. In the juvenile stage, plants appear and clay minerals and organo-mineral colloids form. At the end of this first stage, podzolic tendencies are revealed for leaching effects some migration of sesqui-oxides from the surface. An early juvenile stage exists on 10,000-year-old drift in the north, while on 20,000-year-old drift A+B horizons reach to 30 cm, and developed podzols occur on sands. Thicker podzols occur in a mature stage, with clay formation dominant on the oldest drifts. A senile stage has clay decomposition into Al and Fe oxides and silica; and the soils possess hydrated red or yellow B horizons. Such senile soils characterize the unglaciated areas which have had a weathering time of 0·5–1 million years. On the south-east coastlands there are, in addition, two concentric belts with hydromorphic soils, the one nearer the coast with peaty hydromorphic soils developed on a land surface emerged from a third interglacial transgression (150,000 years old); the other an inland belt, emerged 400,000 years ago, having ground-water podzols and more mature hydromorphic soils. Within each of the five areas mentioned many local variants are found: these are loess, river terraces, eroded areas, calcimorphic and psammitic soils.

Soils in tropical areas exhibit well developed profiles related to pre-Pleistocene influences which may have continued uninterruptedly, but more probably varied between pluvial and sub-pluvial conditions during Quaternary times. While young soils come to resemble much older soils relatively quickly, many laterite crusts remain from Tertiary times in East and in West Africa, especially in presently subhumid areas, and are relic or palæogeographic features. In the Sahel (15°N), transitional between Sahara and Sudan, fossil crusts cover four unconsumed erosion surfaces dating from late Secondary to late Tertiary times, after which time soil formation and landform development are related to alternating pluvial and dry periods and terrace formation.[29] (Fig. 14, 4).

In Uganda[30] two erosion surfaces, a pre-Tertiary Gondwana at 4200 feet (1280 m) and a mid-Tertiary African surface, are both covered by laterite which is now being eroded. However, laterite was not formed over the whole of these surfaces and is less well developed in areas removed from groundwater influence. Near Lake Victoria (2°S) the presence of laterite is explained by variations of water level in the lake in pluvial times and by complete dessication and hardening in mid-Pleistocene interpluvial periods. Thus laterite formed at various times on various low-angle surfaces; the highest plateau laterite in mid-Tertiary times; its lower extensions in end-Tertiary times; while at low level, near the present lake, laterite formed in mid-Pleistocene times as is proved by artefacts. Independently Voute[31] shows that the transgressions of Lake Chad (14°N) correspond to the formation of clayey latosols on terrace and alluvial material, while dry periods had dessication, crusting, erosion and further terrace cutting.

In western and central Australia soils of great age are also found on old land surfaces with low relief, a dry climate and a substrate of resistant rocks. Descriptions of the relation of landforms and soil age in south-west Australia in a system of five zones in the Avon and Swan river valleys (32°N) inland from Perth are worthy of close study.[32]

The time factor is difficult to isolate with certainty. There are problems of intensity, retrogression and inertia to contend with in soil development. But it is infinitely preferable to refer soils to absolute age, or to relative age of site, than to refer to such alleged criteria as 'young', 'mature' or 'old'; for soils may be older or younger than they appear, and such terms, if used at all, should be relative terms, applicable only to a very local area.

Soil is a dynamic three dimensional piece of landscape, synthesizing many present and past environmental influences within its profile. One can have soil landscapes as well as profiles, each with their own form of natural or man-induced responses of vegetation, agriculture or settlement type. The generalized zonal soil group is often quite useless to explain this vastly varying interrelationship of soil and landscape, and the many variants of the zonal or intra-zonal groups are more relevant and helpful to this geographic purpose. However, before the variations of profile morphology within the accepted zonal groups can be studied or understood, and their genesis appreciated, one must consider the dynamic aspects of soil—the processes by which it is differentiated into its diagnostic and interrelated horizons.

1 R. J. Chorley, *Geomorphology and General Systems Theory*, USGS, Prof. Paper, 500-B, 1962

2 E. M. Bridges, Agriculture, 68, 1961, 258–63

3 E. C. J. Mohr and F. A. van Baren, *Tropical Soils*, Hague, 1954. See quotation in H. Jenny, 1941, Table 8, p. 37

4 R. C. Hay, AJS, 258, 1960, 354–68

5 R. Weyl, Neues Jahrb. Geol. u. Pal. Abh., 1954, 49–70

6, 7 See Jenny, 1941, p. 37–8

8 R. Sprague et al., Research Studies, 27, 3, 4, and 28, 1, 1959-60

9 R. L. Crocker et al., J. Ecol., 43, 2, 1955, 427–48

10 L. A. Douglas and J. C. F. Tedrow, SS, 88, 6, 1959, 305–

11 C. O. Tamm et al., Nature, 185, 1960, 706–7
12 E. J. Salisbury, J. Ecol., 13, 1925, 322–8
13 H. Stremme, Zeit. Deut. Geol. Ges., 112, 1960, 299–308
14 G. W. Dimbleby, *The Development of British Heathlands and their Soils*. Oxford Forestry Mem. 23, 1962, p. 37, and V. B. Proudfoot, JSS, 9, 2, 1958, 186–98
15 N. Linnermark, *Podsol och Brunjord*, Lund Inst. Min., Pub. 75, vol. 1–2, 1960, (see vol. 1, p. 195).
16 O. Tamm, *Northern Coniferous Forest Soils* Scrivener, Oxford, 1950, p. 120–1
17 N. Leneuf and G. Aubert, TICSS, 7, 1960, V. 31, 225–8
18 G. Aubert, Bull. Agric. Congo Belge, 40, 1949, 1383–6
19 K. S. Hoeksema and C. H. Edelman, TICSS, 7, 1960, V. 56, 402–5
20 B. E. Butler, *Periodic Phenomena in Landscapes as a Basis for Soil Studies*, CSIRO, Pub. 14, 1959
21 P. H. Walker, JSS, 13, 2, 1962, 178–86
22 R. W. Simonson, AJS, 252, 1954, 705–32
23 R. Lüders, ZPDB, 94, 1961, 47–53
24 O. Fränzle, Eiszeit.u.Gegenw.,11, 1960, 196–205 and Erdkunde, 13, 4, 1959, 289–97
25 V. B. Proudfoot, Nature, 181, 1958, 1287
26 M. J. Gardiner and P. Ryan, Irish J. Agr., 1, 1962, 181–8
27 J. V. J. de C. Cardoso, TICSS, 7, 1960, V. 9, 63–70
28 G. F. Carter and R. L. Pendleton, Geog. Rev. 46, 4, 1956, 488–507
29 J. Dresch and G. Rougerie, Rev. Géom. Dyn. 11, 1960, 49–58
30 J. W. Pallister, CRCI. Geol., XX, Algiers, 21, 1954, 193–9
31 C. Voute, Act. IV Cong. Panafr. de Préhist., 189–207. Reprinted as Tervuren Mus. Sci. Ser. no. 40, 1962
32 M. J. Mulcahy, Zeit. Geom., 5, 3, 1961, 211–25; R. L. Wright, J. Ryl. Soc. W. Aust., 45, 2, 1962, 51–64

8

The process of soil formation

ALTHOUGH the phrase 'soil processes' conjures up such terms as podzolization and laterization, these are *formal processes*—amalgams of many complex processes acting either together or consecutively in the soil to produce a distinct modal profile which, though it may vary in detail, exists mainly within a well defined climatic environment.

Despite the complexity of these formal processes, they are not inseparable. Decalcification precedes podzolization in temperate areas; desilicification comes before laterization in tropical areas. In south-east USA podzolization and laterization act simultaneously in the profiles of red-yellow podzolic soils,[1] though at different depths. Likewise, in subarid areas, calcification and solonization may operate together.

A concept of soil genesis has been outlined[2] as '*an aggregate of many individual physical, chemical and biological processes, all potential contributors to the development of every soil, their rates differing in different environments*'. Each process is regarded as dynamic and irreversible, and they are not in equilibrium. As long as their rates remain in the same proportion, soil development is progressive; as soon as one alters, then the course of soil formation itself alters. The processes combine into two overlapping stages: (1) the formation or accumulation of parent materials, (2) the differentiation of horizons.

Accumulation of parent materials is mainly accomplished by strong physical weathering and superficial biochemical weathering; *differentiation of horizons* is ascribed to additions, removals and transfers of materials and energy (Fig. 1).[3] Most notable *additions* are organic matter and gases; *removals* concern salts and carbonates;

transfers are of humus and sesquioxides. *Transformations* occur too, of primary organic matter to humic acids and of primary minerals to secondary clays, either *in situ* or during transport. All these take place simultaneously in the soil to greater or lesser degree, the balance between them governing the nature of the profile.

All the soils of the world can therefore be viewed as a continuum, with a varied number of properties in common: physical properties such as colour, texture, structure; chemical and biological properties such as clay minerals, biological populations, and relative elemental composition.[2] Thus cyclic changes of short duration resulting from weathering are superimposed on the progressive development of the soil profile.[3]

Normally a complex sequence of physical and chemical 'weathering' processes is involved. Variations of heat and moisture supply lead to varying water intake into the pore spaces and cracks of rocks, and to variable expansion and contraction phenomena in solid rocks. Physical disruption of rocks is achieved by many well documented processes;[4] during erosion, or transportation, debris may be little altered chemically but will be abraded to smaller calibre to form the parent material of regosols. The more fragmented the original rock the more readily is it disrupted, and physical weathering thus increases the specific surface on which chemical processes can act.[5]

The response of mineral fragments to chemical weathering depends either on the cementing agent in sediments and in metamorphic rocks, or on the resistance of the crystal lattice of igneous rocks to chemical activity. The more complex or readily soluble minerals there are in the rock, the more readily it succumbs to physical weathering, combined with simple solution and incipient chemical weathering.

Physical weathering of granite may give sandy pervious materials, while later chemical breakdown produces a clayey moisture-holding residue, thus increasing hydration and hydrolysis.[6] In mountain and arid regions physical weathering is intense, chemical alteration and bacterial activity are minimal; soils therefore lack clay and humus as binding agents and are highly erosive.[7]

In the opening chapter the role of organisms in the earliest weathering stages was stressed. Polynov[8] doubts the possibility of 'sterile weathering', even in deserts—'there is no isolated mechanical weathering in nature which is not associated with chemical changes'. However, soils derived from gneiss at height on Trollheim, Norway,[9] show no chemical weathering, while in the high Sierra Nevada, Spain, mechanical agencies were far more effective than the low rainfall in disrupting a laminar calcareous metamorphosed clay to form soil, which clay at lower levels is weathered entirely by solution.[10]

The appearance of felspar grains is a good indication of type of weathering.[11] If they are fresh then physical weathering dominates, if rounded or soft then chemical weathering is significant. Alkaline felspars are little modified by chemical weathering until finely divided by physical forces; though coarse-grained Ca- and Na-felspars are readily altered by hydration and, like biotite, are then transformed into clay on losing bases.[9]

Thus mineral topsoils subject to strong physical forces, but weakly altered chemically, have element and mineral contents closely related to the parent rock, the only elements removed being those most readily soluble or oxidizable. Yet, in chemically strongly weathered topsoils, many less readily soluble substances are also lost and primary minerals are replaced at depth by secondary minerals with distinctive chemical compositions—the result of condensation and resynthesis from the percolating soil solution.[12] In humid tropical areas, for example, S, Ca, Na, Mg, K and Ba are all removed; the contents of some other elements are the same as in the parent rock—Cr, Ni, Co, Cu and Mn; while the concentration of Fe, Ti, Mo, V and Al is increased (relatively) in the upper part of the soil.[13] (See Fig. 3.)

Hence chemical weathering is both destructive and constructive. It is destructive of solid rock and of primary and of secondary minerals in the upper soil; it is constructive of secondary minerals formed from decomposition products chiefly in the subsoils of developed soils.

At least six fundamental processes are involved in chemical weathering—oxidation, hydration, hydrolysis, carbonation, chelation and ion exchange. The first three are atmospheric in origin; the fourth and fifth are biochemical, the sixth an internal chemical reaction of minerals and the soil solution.[14]

Oxidation occurs when constituent atoms of a mineral lose electrons to a weathering agent. Normally the elemental oxygen combines with the weathering substance. Few minerals are immune, and the products of the reaction—conditioned by the presence of moisture—are usually soluble. Atmospheric oxidation of pyrites, with water vapour present, is well known:

$$2FeS_2 + 7O_2 + 2H_2O \rightarrow 2FeSO_4 + 2H_2SO_4$$

Oxygen need not be involved in the loss of electrons, for bacteria may break down organic matter to ammonia, which is oxidized to nitrate and then reduced to toxic nitrites. Iron can also 'oxidize' to FeS_2; the sulphur being derived from organic matter, from atmospheric pollution, or from volcanic sources, forming a hydride H_2S—twice as soluble in water as CO_2—which easily precipitates iron from solution to form iron sulphate.

Continuous *reducing conditions* in connection with slow water movement and leaching increases the solubility and removal of Fe.[15] Soils in a reduced state liberate large amounts of exchangeable cations—Fe, Mn and Al—and the pH of waterlogged soils becomes very low. Ferrous salts are oxidized to ferric by oxygen in neutral solutions : $4Fe^{++} + O_2 + 2H_2O \rightarrow 4Fe^{+++} + 4OH^-$; ferric salts are reduced by hydration in acid solutions or by H_2S or SO_2, thus

$$2Fe^{+++} + 2H_2O + SO_2^- \rightarrow 2Fe^{++} + SO_4^{--} + 4H^+$$

Oxidation therefore increases the valency of the mineral and lowers that of the weathering agent; reduction leads to a decrease of valency of the mineral and increases that of the agent, as well as the total negative charge.

The terms oxidation and reduction are therefore relative, for both may occur in any soil. In wet conditions reduction prevails and the soil quickly uses up any introduced oxygen; in dry conditions, oxidation prevails. The boundary between oxidation and reduction is often narrow, for they are complementary, the balance changing seasonally, and one cannot take place without the other.[15]

The transfer of electrons during oxidation-reduction may be likened to an electric current and can be expressed in millivolts, and the strength of this reaction expressed as the *emf*, the *redox potential* or *Eh*. Normally the greater the negative potential of a metal in a mineral, the easier will it pass into an ionic state. Potash (K) has the greatest negative potential (-3v) and the order of increasing, or less negative, potential is Ba, Ca, Na, Mg, Al, Mn, Fe, Ti, Co, Ni, Pb to 2H (which is zero) and for Cu$+0\cdot337$ v.[16] Any metal will displace any of those below it in the list from solutions of its salt; the more negative it is the more powerful a reductant it is, or, alternatively, the greater its capacity to suffer oxidation.

The Eh of a soil may be considered as an average of all the oxidation and reduction sites operating within it. The *redox potential* of a long waterlogged organic soil is c -700 mv. A temporarily flooded organic soil with O_2-rich water has c -460 mv. In a permanently saturated mineral soil Eh is c -230 at the surface, -130 mv at depth. Positive figures of Eh show oxidation with $+530$ mv in a dark chestnut soil, 400 mv in a clayey soil on till. The Eh of aerated soils is lowered by organic matter and by microbial activity. Eh is thus a far more reliable indicator of drainage state than colour; for not all soils show the blue-grey colours typical of reduction when wet, especially calcareous soils.[15]

Hydration is the process which combines water with rock constituents and is most important in warm humid regions. Many silicates, oxides, carbonates and sulphates are affected, forming hydrous compounds—eg gibbsite, $Al_2O_3.3H_2O$, and limonite. Some minerals are hydrated while still under pressure in the rock matrix and are further hydrated and 'slake' on exposure, softening and increasing their bulk. Apparently fresh, such rocks disintegrate at a touch. Hydrated minerals lose their lustre and hardness, while swelling hydrated montmorin clays exert pressures of up to 10 t/ sq ft. *Dehydration* also occurs; hydrous ferric oxides change from yellow to red on drying, while dehydration of clays gives shrinkage of up to 25% of the volume of wetted materials. Hydration occurs along with hydrolysis, oxidation and carbonation.

Hydrolysis is the reaction of dissociated H^+ and OH^- ions of water with the ions of mineral elements. The OH^- ions combine with metal cations (eg K^+) which move in solution to the ocean as salt; while H^+ ions combine with aluminosilicate anions to form difficultly-soluble clay minerals or acid silicates. Hydrolysis is enhanced by repeated washings of rapidly percolating water leaching away dissolved products and re-exposing particle faces for further reactions. The products of hydrolysis may also be removed by plants, by chelation, or by colloidal adsorption.

H^+ ions are very small, highly charged in relation to their size, and readily replace other cations. There are many sources of H^+ ions in nature—rainwater, snow, carbonic acid, mineral acids in the soil, plant roots and soil organic and humic acids. The pH of the soil—or of its solution—is the log of the reciprocal of the H-ion concentration and is slightly at variance with what it is supposed to measure—the H-ion activity. Acidity is then an excess of H ions; alkalinity an excess of OH ions. The pH of pure water is 7, having one ten-millionth gram of H-ion per litre, and pH 3 expresses an H-ion concentration of 0·001 g/l. The many sources of H-ions render the pH highly variable in the profile, depending on the 'freshness' of the reacting water. On a global scale, soils with pH <6 are common in cold temperate areas (eg Finland), where there are few soils with pH >8; but the reverse is true in arid lands (Egypt) where 40% of soils have pH >8 and few, if any, have pH <6.[17]

Carbonation is the combination of carbonate (CO_3^{--}) or bicarbonate (HCO_3^-) ions with a mineral. It may be illustrated by reference to a lime-rich till colonized by plants.[18] The plant debris is broken down by organisms and mixed with the mineral matter to form a mull horizon. When this is washed with rainwater this reaction occurs:

$$3H_2O + 3CO_2 \rightarrow H_2CO_3 + 3H^+ + CO_3^{--} + HCO_3^-$$

which is followed by:

$$CaCO_3 + H_2O + nCO_2 \rightarrow Ca^{++} + 2HCO_3^- + (n-1)CO_2.$$

Soluble bicarbonates are formed (ie acid carbonates), and these solutions are more potent solvents than their relatively weak acidity would indicate. In the atmosphere there is only 0·03% CO_2 by volume; in rainwater 0·45%, and in surface organic layers c 1·0%. Water in equilibrium with the atmosphere thus has a pH of 5·72; with soil air, pH 4·95.[19]

For the release of $CaCO_3$ from the till an excess of CO_2 is required. As the free CO_2 content of water is conditioned by the partial pressure of CO_2 in the adjacent atmosphere, this condition determines the rate and quantity of $CaCO_3$ release. If the CO_2 pressure in the soil water should fall, $CaCO_3$ is redeposited; yet, if insufficient $CaCO_3$ is released to achieve equilibrium, the leaching soil water will remain acid, and the reaction remain below neutral even though $CaCO_3$ is being released. The solubility of $CaCO_3$ decreases with increasing temperature at constant CO_2 pressure; thus at 30°C only 0·052 g/l are removed, but more at 0°C—0·081 g/l. When lime has been completely removed from the upper layers of till, biological activity changes and mor humus forms.[20] Yet at depth there is some point where lime is still present and a marked change of pH occurs. This is the *acid front*,[18] which moves more rapidly to depth in permeable soils and sands in high rainfall areas and on hillocks, than on fine-grained soils or in depressions.

Once lime has been removed, acid carbonate water is very active in decaying other minerals, especially Na-, Mg- and Fe-bearing silicates and Mn and Fe oxides:

$$R_2SiO_3 + CO_2 + H_2O \rightarrow RCO_3 + SiO_2 + H_2O$$

In mull soils Fe and Mn occur as difficultly soluble ferri- and mangani-compounds, but replacement by mor gives reducing conditions and ferro-compounds form, which are easily dissolved in CO_2-rich water. Yet Al is not as easily reduced as is Fe above pH 4·5, for the $CaCO_3$–CO_2 equilibrium acts as a buffer and Al hydroxides are precipitated only at pH 3 to 4·4.[18]

The next step is to *chelation* or 'complexing'.[21] Chelating agents are amino-acids or else complex ring-structured organic compounds produced by lichens, by mycorrhiza or else released by roots. All have the ability to combine with metallic cations in the soil, a weak acid proton of the chelating agent (H^+) being displaced by the metal ion. Experiments using leaching water alone on mineral soils in which chelation is known to occur had very little effect in mobilizing the metals, such as Fe, Al, Mg and Si, which can be moved most strongly by chelation.[22] Yet leaching with water containing extracts from leaves and forest litter did dissolve sesquioxides, the extracts holding the Fe and Al ions—removed from the primary minerals or hydroxides—on the inside of their ring structures by valence bonds. Such organo-metallic chelates are stable in an acid medium, and do not combine with other material in solution. Only when they have reached an illuvial zone with pH > 5 is their stability disturbed and the metal redeposited.[23]

If we now return to our leaching solution, ions pass spontaneously from a solid silicate mineral to it and are gradually decomposed by this solution process and the products are eluviated. Hence the process is known as *soluviation*.[24] However, if the leachate is rich in chelating agents, the mineral decomposes in a different and more intense way, for the agents stabilize the dissolved ions to a far greater degree than their solubility in water without such agents. The combination of chelation and subsequent eluviation has been termed *cheluviation* and removes Fe and Al more readily than Si. Soluviation, with removal of Ca, Mg, Na and K by ion exchange and their redeposition at depth is characteristic of pedocals; accumulation of Fe and Al oxides at depth is characteristic of cheluvation and of pedalfers, especially of podzols.[24]

Ion exchange.[25] Virtually all soil particles are surrounded by a film of water, the molecules of which are polar (H^+OH^-). The surface of the mineral particle has negative charges which attract the H^+ cation and the H_2O molecules are regularly oriented, the water having solid properties. In the clay fraction of the soil this adsorbed water layer usually contains other cations which have migrated from the liquid; it is c 0.005 μ thick, only 1/500th part of the breadth of the clay particle. The liquid property is not fully established for c 0.1 μ from the interface.

Ion exchange is thus the exchange of an ion held by a negative charge near the mineral surface with another present in the contacting liquid or electrolyte. The exchange may take the form Na-clay$+H^+\rightleftharpoons H$-clay$+Na^+$; though exchange also occurs between adjacent mineral surfaces or between roots and a mineral surface.[26] Other cations taking part in this migration and exchange are Ca, K, Mg, H and Al. Anion exchange, by Cl, NO_3, PO_4, SO_3, or SiO_4, also occurs.

Soil colloidal molecules, composed of two layers of charges— the inner negatively-charged layer of acidoids surrounded by cations (Ca^{++}, Mg^{++}, K^+, NH_4^+ or Al)—may be in an electrically neutral state, positive charges counter-balancing the anions at the isoelectric point.[27] However, if the electrolyte has excess KCl (for example), the K ions replace the Ca ions of the 'solid' and H ions may then replace the added K, forming $CaCl_2$ and KCl by successive ion exchange.[28] The released cations combine with free anions to form acids, alkalies and salts which are dissolved and leached away as HCl, H_2CO_3, $CaCl_2$, H_2SO_4 and others, while slightly less soluble compounds such as $CaCO_3$ or $FePO_4.2H_2O$ remain in the soil for a slightly longer time.[29]

The capacity of the soil colloids to exchange base ions for those in the soil solution is known as *cation exchange capacity* (CEC). It is defined as the amount of exchangeable cations expressed as milliequivalents per 100 grams (me gm) of clay determined at pH 7, and it is a measure of potential fertility, being related to the clay and organic matter contents of the soil and dependent, too, on the structure and composition of the clays, and on the environment.

Polynov[30] expressed an order of the relative mobility of elements during weathering by comparing the element content of rocks, soils, river and sea waters; tracing the history of elements from their parent rock mineral through secondary materials on land to a final ionic state in sea water. He found the following order and phases of mobility, since refined by later workers:[31]

TABLE 9

PHASE MOBILITY OF ELEMENTS DURING WEATHERING[30]

Phase I $Cl^-=100\%$; $SO_4^{--}=57$
Phase II $Ca^{++}=3$; $Na^+=2\cdot4$; $Mg^{++}=1\cdot3$; $K^+=1\cdot25$; F.
Phase III SiO_2 (colloidal)$=0\cdot2$; P; Mn.
Phase IV $Fe_2O_3=0\cdot04$; $Al_2O_3=0\cdot02$; TiO_2

For Polynov's scheme the most mobile elements of phase I are removed from the parent material and soil first, though not necessarily to exhaustion before those of phase II begin to be removed. Normally the lower the phase the more evenly the elements move from the upper soil layers, often to the drainage waters without redeposition in the subsoil.

In natural environments various cations are therefore either dominant or absent in percolating waters.[31,32] Where seeping water is acid (pH 4·5–5·5), soil clays tend to become H-clays; if neutral, then OH clays form; if alkaline, clays are dominated by Na, Ca and Mg and pedocals form, for the reaction between minerals and soil water is less intense (Table 10).

The composition of clays therefore varies in complexity depending on the internal weathering environment; the more acid and leaching this is, the fewer the elements available for resynthesis of clays. Moist warm to cool climates usually have low clay content of a varied and complex kind,[33] while wet hot climates have much higher percentages of simpler clays such as kaolinite.[34] For example, basic igneous rocks at 18°CØ have soils with 50% kaolinite clay, at 10°CØ only 10–15% of varied clays.

If rainfall is low, drainage deficient, and evaporation high, only carbonates and bases are removed from the soil surface and the remaining alkali-earths, silica and sesquioxides form illite or montmorillonite. With marked rainfall deficiency or concentrated soil solutions, chlorites may form. If rainfall increases (though with dry periods) and drainage is good; then pH is low and Al_2O_3 and SiO_2 remain and combine to form kaolin clays. Ferric iron cannot enter into the crystal lattice of kaolin clay and remains free, in oxidized form, to give the intense red colours of tropical soils. If rainfall is profuse, however, in ill drained areas—on tropical coasts, in moss forests or in equatorial areas—then the two minerals of kaolinite are separated as bauxite ($Al_2O_3.2H_2O$) and soluble silica.[35]

A close relationship also exists between the acidity of the soil solution and the stability of the clay minerals, most of which begin to break up at pH 5·5, rapidly so at 4·5,[36] and leaching frequency also has an effect.

TABLE 10

COMPOSITION, CEC AND ENVIRONMENT OF SOME CLAY MINERALS[37]

Clay	pH	Environment	Formation	CEC (me/100 g)
Illite $K_n(Al_4Fe_4Mg_{10})(Si_8nAl)O_{20}(OH)_4$	—	Temperate soils, podzols and g/b podzolics,shales, tundra soils	Slight leaching (=hydromica)	10-40
Chlorite $(Mg,Fe)_5Al(AlSi_3)O_{10}(OH)_8$	7	Developed aridisols	Stable in alkaline conditions	10–40
Montmorillonite $Al_4Si_8O_{20}(OH)_4 \cdot nH_2O$	7	Neutral conditions, chestnut and prairie soils, moist gleys and margalitic soils	Unstable under leaching	60-150
Kaolinite $Al_4Si_4O_{10}(OH)_8$	4	Acid tropical soils R/y podzolics	Leaching and oxidation	3–15

D

Jackson and others[38] have formulated a sequence of increasing order of weathering intensity related to those minerals of a primary or secondary nature most noticeable in the soil—gypsum, calcite, hornblende, biotite, albite, quartz, illite, montmorillonite, kaolinite, gibbsite, hematite and anatase. These successively dominant minerals are considered the resultants of factors of weathering intensity (temperature, moisture transfer, acidity and *redox*) and of factors of magnitude (size of clay particles and the nature of the clay mineral, especially its CEC). Other factors are time (the length of the reaction) and the relative position in the soil profile. The uppermost part of the soil has more and fresher waves of leaching penetrating downwards and, by the time the subsoil is reached, this fresh water will have been greatly altered by ion exchange and other processes of chemical weathering.

Weathering is often most intense in the lower A2 and upper B2 horizons of pedalfers, while upper A horizons derive some bases from plant ash. Yet A horizons lose elements of phase IV on cheluviation which are redeposited in the upper B horizons; C horizons receive minerals from the B2 layers but of a lower phase in Polynov's scheme. In pedocals, B layers receive minerals of phases I and II for the elements of higher phases are only slightly mobilized. In pedalfers, if conditions are neutral or only slightly acid, clay particles are leached mechanically and are not broken down, a phenomenon known as *lessivage*, the clay being redeposited as clay skins on ped surfaces (see 7A, Figs. 9—11). If conditions are acid, clays are broken down and dissolved, their constituents moving by cheluviation and being redeposited and re-synthesized at depth.

The marine clays of western Europe exemplify the first stages of Jackson's sequence. Raised above sea level by isostatic uplift only 2000–3000 years ago, percolating water is actively leaching NaCl from them. In 8000-year-old cultivated clays Na is still being removed, but lime is being leached at a higher rate.

The stage of *calcite* removal is best illustrated by the carbonate (caliche) horizons of subarid areas at shallow depth. The next stages of weathering are the modification of the crystal lattice of primary minerals containing Na, Ca, K and Mg to form hydrous mica and *illite*. This is seen in tundra soils and on young tills *c* 10,000 years old. In till near Hudson Bay, *hydromica* dominates the clay complex, while on older tills (*c* 20,000 BP) in Komi ASSR, amorphous clays and *montmorillonite* dominate the 1μ fraction. On tills of similar age in Manitoba, montmorillonite and illite are present with some chlorite. Slightly altered felspars are also common at this stage. In further intermediate stages traces of kaolinite appear, and are common, along with fine quartz, in the granite-derived soils of the southern Urals, in long-developed desert soils and the grey or brown forest soils of subhumid regions. In podzols, *illite* and *kaolinite* occur, and free hydroxides represent the 'early advanced' stage of weathering in materials $>50,000$ years old, or in younger, shallower, soil horizons which have been very intensely altered.

Very little mica is present in soils in areas of more advanced weathering, as in India, where halloysite, hydrous hematite and *goethite*—FeO(OH)—are dominant. The final stages of weathering are dominated by *gibbsite* and *hematite*, as in old tropical clays; while A layers of intensively weathered tropical soils comprise up to 25% TiO_2 as *anatase*, together with limonite or hematite.

Thus to quote Jackson[38] 'the weathering stage of the soil colloid tends to advance with increasing proximity to the surface . . . the stage is also expressed geographically for weathering intensity is controlled by the geographic distribution of climate together with time, and with biotic activity—which is also large controlled by climate'.

1 R. D. Krebs and J. C. F. Tedrow, SS, 85, 1, 1958, 28–37
2 R. W. Simonson, PSSSA, 23, 2, 1959, 152–6
3 A. A. Rode, *The Soil-Forming Process and Soil Formation*, Moscow, 1947, IPST, 1962
4 P. Reiche, *A Survey of Weathering Processes and Products*, Univ. New Mexico, Pub. Geol. 3, 1950
5 M. L. Jackson et al., *Adv. Agron.*, 1953, V, 211–319
6 D. Collier, Ann. Agron., 12(3), 1961, 273–331
7 J. L. Retzer, PSSSA, 13, 1948, 446–8 and SS, 96, 1963, 68–74
8 B. B. Polynov, transl. in S and F, 14, 1951, 95–101
9 H. Holtedahl, N. Geol. Tids., 32, 1953, 191–226
10 E. G. Rios and A. M. M. Ortega, An. Edafol. Fisiol. Veg., 9, 1950, 475–536
11 O. Tamm, Medd. Stat. Skogsförsök., 25, 1, 1929, 1–28
12 A. N. Puri, Colloid Chem., 7, 1950, 443–53
13 Mohr and van Baren, op. cit., ch. IV

14 W. D. Keller, *Principles of Chemical Weathering*, Univ. Missouri Press, 1957

15 W. H. Pearsall, Emp. J. Expt. Agric., 18, 1950, 289–98

16 See the standard texts on Physical Chemistry

17 O. Arrhenius, Ark. Botanik, 1922, 18, 1, 1–54

18 W. Christensen, Medd. Dansk. Geol. För., 15, 1, 1962, 112–22

19 A. J. Ellis, AJS, 257, 1959, 354–65. See also J. D. Hem, USGS Water Supply Paper, 1459-B, 1960

20 W. Laatsch, *Dynamik der Mitteleuropäischen Mineralböden*, Dresden, 1957

21 There are many references to this process: PSSSA, 27, 2, 1963, 179–86; Ann. Rev. Plant Physiol., 14, 1963, 295–310 and ref. 14, 33–4

22 C. Bloomfield, J. Food and Agric. Sci., 8, 1957, 389–92, and L. N. Alexsandrova, Poch., 11, 1954, 14–29

23 F. E. Broadbent and J. B. Ott, SS, 83, 6, 1957, 419–27

24 L. R. Swindale and M. L. Jackson, TICSS, 6, 1956, V. 37, 233–39

25 D. Carroll, BGSA, 70, 1959, 749–80

26 W. D. Keller and A. F. Frederickson, AJS, 250, 1952, 594–608

27 G. H. Cashen, Chem. and Ind., 43, 1961, 1732–7

28 S. Ståhlberg, Acta Agric. Scand. IX, 3 and X, 3, 1959–60

29 P. G. H. Boswell, *Muddy Sediments*, Heffer, 1961, pp. 125

30 B. B. Polynov, *Cycle of Weathering*, Murby, 1937, p. 163

31 A. I. Perelman, Doklady, 103, 1955, 669–72

32 R. E. Grim, *Clay Mineralogy*, McGraw-Hill, 1953

33 H. W. van der Marel, SS, 67, 1949, 193–207

34 T. Tanada, JSS, 2, 1, 1951, 83–96

35 J. P. Bakker, Zeit. Geom., Supp. 1, 1960, 69–92

36 C. Bloomfield, Nature, 172, 1953, 958; JSS, 4, 1, 1953, 5–16, and 5, 1, 1954, 50–6

37 D. Carroll, in *Sedimentary Petrography* (Milner), 1962, chapter III. Also R. W. Grim, ref. 32, et al.

38 M. L. Jackson et al., J. Phys. and Colloid. Chem., 52, 1948, 1237–60 and PSSSA, 16, 1952, 3–6

9

Soil description, classification and nomenclature

> Most systematicians stand in the same relation to
> their systems as a man who builds a great palace and
> lives in a barn adjoining it. SØREN KIERKEGAARD

TO MANY, no doubt, this (the title) is the point at which soil geography begins. Yet one cannot hope to understand the morphology, classification and distribution of soils without a knowledge of their environment and genesis.

In classifying soils one arranges them into assemblages according to selected characteristic properties. There are several ways of doing this: some are more relevant to geography; others, equally valid, to agronomy or engineering.[1] Properties such as texture, base content or shear strength, proved significant for some local practical purpose, are used in such *empirical* schemes. Another approach, the *morphologic* approach, classifies soils on the basis of their profile as noted in the field, taken as evidence of their formative process and stage of formation. A third approach, the *genetic*, attempts to explain and classify the profile from known formative or environmental factors. It is of course possible to combine these approaches, as in the Seventh Approximation,[2] in which measurable horizon properties and genetic factors of proven relevance to soil formation have been used to define the aims of the morphological approach.

The genetic approach considers soil as a sequence of horizons, an isotropic body, occupying a site and related to geomorphic, climatic, biotic and other influences of a zonal type.[3] The system derives from Dokuchayev and involves field study of profiles, laboratory analysis of properties, and study of data on the area compiled by other sciences.

Yet soils are not classified from a general concept of a zonal group divisible into many lesser variants. The basic concept underlying classification is an upward and integrative one, moving from a single profile towards integrated assemblages; first by sorting them into comparable or slightly variable *series*, which are then amalgamated with increasingly broad scope—associations, families, groups and orders. A single profile is rarely unique, but recurs on similar sites,[4] and the higher categories[5] (eg groups) recur too, in similar areas of a regional or sub-continental scale. Thus, because of the different levels of categorization, soil descriptions can be made for local and for regional surveys and the data and terms of the one should fit into the scheme of the other.[6]

Soil classification by reference to the profile has often been subjective, for the processes of formation are not always readily inferred from profile morphology. The most striking of all soil properties is colour, which has been used at all scales of characterization and classification—individual grains have colours inherent to their mineral type or else derived from a coating during soil formation. Soil horizons have their own diagnostic colours; while many soil groups have names related to their dominant colour, which has caused much confusion in soil terminology and classification.

The main soil colours are red, yellow, brown, black and white, and their intergrades 'orange', chocolate and grey. Some colours are derived from inorganic, some from organic components. In the past colours were characterized subjectively; now, fortunately, it can be defined objectively, using a descriptive term and code number on the *Munsell Soil Color Chart*,[7] at a defined state of wetness.[8] In the past, soils of temperate areas were referred to as brown (forest) soils, or as brown earths, a term also applicable to a soil of semi-desert areas (arid brown soils), though this brown is mainly lighter in colour. Similarly the term 'chestnut soil,' or *kastanozem*, is derived from the similarity of colour between this subhumid Russian soil and the bark of the Spanish chestnut, whereas to a British student the term has the connotation of 'dark reddish brown', rather than 'greyish brown'.

Many popular Russian words, now in widespread scientific usage, refer to colour, sometimes directly as in *krasnozem* (red soil) and *chernozem* (black soil) or, as in *podzol*, by implication. 'Chernozem' in the early Russian literature referred to the organic layers of any soil[9] (just as 'podzol' referred solely to the bleached horizon of any soil[10]) before these terms were limited to the present soil group or total profile by pedologists.

Wilde has rejected the 'terminological spectrum' with its confusion and duplicity,[11] but one must acknowledge that these soil names are meaningful to the layman as well as being diagnostic of certain processes of soil formation. Many colour terms are worthy of some respect for they were actually coined by the people using these soils, and it is the pedologist who has created confusion by inventing some and mistranslating others, as well as inadequately correlating most.

The use of more objective criteria as a means of classifying soils such as the ratio of colloidal silica to sesquioxides in the weathering complex was in vogue before 1939.[12] Now that widely differing complexes are known to have similar ratios this has been abandoned as a basis of classification, though it showed the way to intensive chemical analysis as a reliable way of establishing the actual mode of soil formation. Other classifications referred to internal properties such as the nature of the dominant exchangeable cation. Gedroits[13] showed that soils with the H^+ ion dominant were podzols; joint H^+ and Ca^{++} dominance gave brown forest soils; rendzina or chernozem were dominated by Ca; saline soils by Na. Others considered that if the adsorbing complex was intact, then chernozem would form; that podzols formed if it was partially destroyed; and complete destruction was typical of tropical soils.

Both Robinson[14] and Marbut[15] related soils to their degree of leaching, and had two major categories (1) pedalfers and (2) pedocals. A lime accumulation horizon was present in the second and each was divisible into lower categories, for example, siallitic and allitic weathering in the pedalfers. Intrazonal and azonal soils were regarded as soils of a 'lower category' than zonal soils, reflecting Marbut's emphasis on the fully developed 'mature' soil.[16]

Schemes of soil classification based on climatic indices generally find favour with geographers and are valuable in stressing the differing pedogenetic intensities of hydro-thermal regimes.[17] It is then possible to arrange zonal soils as an overlapping sequence related to varying weathering intensity (Fig. 4). In contrast Jenny's total factorial approach[18] is more suited to the explanation of local soil differences at a lower level of classification, where it is preferable to study a soil profile and infer the formative factors, which is the basis of the morphological approach, than to infer soil properties from a knowledge of the assumed causal factors of climate—the genetic approach.

Horizon description and categorization. The use of capital letters—ABC—to categorize soil horizons was introduced by Dokuchayev in 1883 (see Fig. 2); and their subdivision by numbers introduced perhaps in 1900. Systems of categorization vary in different countries and are by now very complex.[2,4,19,20] The superficial organic horizons pose several problems. Best referred to as L (litter); F (fermentation) and H (humus) layers, they are given a separate depth scale from n to O cm—from the surface to the organic-mineral interface. Each horizon is subdivided (F1, F2) if need be. Soils with mull horizons may be characterized as mineral horizons, beginning at O to n cm, and termed Aml or Ao1, Ao2 horizons.

Mineral horizons are measured separately and cumulatively (0–10 cm A; 10–30 cm B; 30–70 cm C, for example), and the upper eluviated mineral layers with most organic matter, and possibly least clay, are called A layers. Often darkened in their upper part, they may have subhorizons termed Ah (humic staining), followed at depth by A2 and A3 horizons. Consecutive numbering rarely goes beyond '3', for minor differences may be categorized by letters (or numbers), as in A21, A22. Other variants are Ap for ploughed surface horizons, and Au for disturbed or washed-in surface horizons on slopes or in hollows. The term Ae for eluviated horizons is increasingly used.

In well drained, developed soils an illuviated subsoil is found—the B horizon—in which one or more of bases, clays, sesquioxides and colloidal organic matter may accumulate, derived from the A horizons. The B horizons are therefore finer textured, more compact and of more developed structure than the A horizons. They are also divided consecutively downwards by numerical suffixes (B1, 2, 3) but more lower case letters are used—Bh (humic),

Bs (sesquioxide), Bg (gleyed), Bca (calcic), Bt (clay-enriched) and Bsa (saline).

The C horizon, the presumed parent material or substrate, resembles the parent rock, or is a more heterogeneous fragmented material if derived from transport or mass movement. The C horizon is but little affected by organisms; it is richest in secondary clay minerals in tropical areas, though the B horizon is usually richest in clay in temperate areas. C horizons also receive materials leached from A and B horizons, especially in subhumid areas. Deeper parts of C horizons, or slightly altered rock, are known respectively as CR, or R horizons.

G horizons are permanently water-logged substrates, subjected to gleying; laterite horizons are referred to as BL horizons and crusts as Lcr—for there is little chance of confusion with organic L layers. Separate Ca, Sa, and T (ie peat) horizons are possible. A 'D' horizon may be found at the base of a profile which, though it cannot be considered as a genetic parent material, does affect soil formation by its presence. An example would be a layer of clayey till underlying loess, in which a brunizem is developing.

All horizons do not occur in all soils, one can have, for example, AC or A(B)C profiles; or LF, BC profiles on a revegetated eroded soil. A letter or number in brackets (B) or A(h) indicates that the horizon is but weakly developed. Transitional horizons may be written B/C, while buried horizons may be written ABC, Ab,Bb[4] or (better) ABC, IIA, IIB.[2] Recognition of distinctive horizons and diagnostic sequences is the key to soil classification, and the description of single profiles is achieved by measurement of depth and properties of subhorizons.[19] (See Fig. 5.)

Categories of soil classification. From the level of the individual *soil profile*—a two-dimensional face—one proceeds to a rigorously described and delimited soil body, a three-dimensional entity termed a *pedon*, which is defined as 'a volume of soil including the full solum to within the underlying unconsolidated rock and whose lateral cross-section is hexagonal or circular in shape and about 1 to 10 sq.m in size'.[21] One then proceeds through several higher categories known as soil phase, type, series, association, family, group, suborder and order. The higher the category, the fewer the precise statements that can be made concerning it.[5]

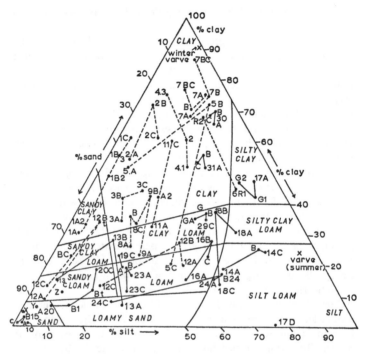

Fig. 5 Textural profiles of soil groups and other materials. The familiar textural triangle, standardised to the International scale, reveals many aspects of soil character and horizon formation. The soils are numbered and their profile formula is given:

1A–1C; humic ferrallisol, Congo
2ABC; ferrallisol on granodiorite, Angola
3ABC; ditto on slate
4, 1–3; rotlehm, Cyprus
5ABC; rubrozem, S. Brazil
6R1–2; relic terra rossa, Cyprus
7A,B,BC; black tropical soils, São Tomé
8A,B8C; solonets, and
9A,B8C; solodized solonets, N. Kenya

10; tropical podzol, A2
11A–C; lithosolic on basalt, São Tomé
12ABC; red podzolic on granite, S. Pennsylvania
12A,Bt,BC; red podzolic on coastal alluvium, N. Carolina
13ABC; grey-brown podzolic, Illinois
14ABC; sol brun acide with fragipan on brickearth, Belgium

15ABC; podzol on fluvioglacial sand, W. Jylland
16ABC; fragipan on loess, Illinois
17AG1G2; calcgley, Banff, on Jurassic clay till (D)
18ABC; brunizem on loess, Illinois
A19C; southern brown earth, Guadalquivir
A20B1BtC; brown podzolic, Warwick

X,Y,Z; A2 layers of ultra-podzol, iron and iron-humus podzols, Matlock
23ABC; podzol on shaley till, Banff
24ABC; arctic loess on alluvium
29GA,GB,C; humic gley, Indiana
30ABC; marine clay, S. Carolina
31ABC; ditto, S.W. Jylland

Also note varves (X)

The fundamental soil mapping unit referring to the profile is the *series*.[22] Defined as 'soils similar in every respect, except for their surface texture', 'they are developed from the same or similar parent material and have the same range of colour, structure, horizon sequence and the same conditions of relief and drainage'—hence a similar site, origin and mode of formation. Avery states that the series should be regarded primarily as a geographic concept,[23] 'a landscape unit with a limited and definite range of soil profile variation based on a limited range of parent materials'. A series is usually given a geographic name, from the area in which it was first described, and a textural designation may be added,[4] for example, 'Monona silt loam', or 'Icknield series'.

Series are assembled with other soils of characteristic geographic proximity into soil *associations*,[24] and they may differ greatly, though side by side, if parent materials differ, or slightly as a result of changes of drainage condition on adjacent sites.

Soil type is conventionally explained[4] as textural variations of the A horizon within a soil series. Several textural variants are assembled to form a series, each type usually related to slight changes of slope. A *soil phase* is a further variant, dependent on erosional state, expressing variations of depth or stoniness, or disruption of the A horizons. Phase is the lowest unit of classification and has the narrowest range of characteristics (eg Icknield series, shallow phase).

Another unit of soil mapping, though not of classification, is the *complex*[19] where several series may be too intermixed to be clearly plotted on the scale of a published map; while a number of distinctive soils, occurring in close but variable proximity and association, without constant relation to micro-landform, (eg in alluvial, till or desert areas), are termed *soil mosaics*.[25]

Soil families—the next higher category to the association—are less well defined and are best considered as subdivisions of groups,[24] differing, for example, in the structure, colour, or degree of mottling of the B horizon. The podzol group, for example, may have family variants (i) with thick, soft, Bs horizons or (ii) thin, hard, Bs (pan) horizons; while chernozems may be 'leached', 'compact' or 'typical'.

I shall leave the group for the time being and now examine the three *orders*. *Zonal soils* are defined as 'those soils occurring over large areas (or zones) limited by geographical characteristics; having well developed properties; sited on well drained undulating land; on parent material which has been in place long enough for the active forces of climate and organisms to have expressed their full influence'.[4]

Azonal soils lack a well developed genetic profile either because of youth, or because parent material or relief prevent the development of a characteristic profile. Each may be associated with any zonal group. Azonal soils have no suborders, and there are three groups—*lithosols* developed on solid rocks; *regosols* on unconsolidated materials; and *alluvial soils* on active flood plains.

The third *intrazonal* order has soils which reflect the dominating influence of some local factor of relief or parent material over the zonal effects of climate and vegetation. There are three suborders—*halomorphic*, *hydromorphic* and *calcimorphic*, each divisible into several groups.

The zonal order was divided, or rather re-organized, in 1949[5] into six *suborders* according to zonal climate and vegetation, each divisible into several great soil groups. The suborders, each with their most widespread groups, are known as: (1) *tundra, subarctic or polar soils*, including arctic brown soils, peaty, gleyed and cryogenic soils; (2) *light-coloured surface soils of cool temperate areas*—the *podzol* and its variants as well as the incipient or 'podzolic' forms such as grey wooded soils, grey, brown, grey-brown, and sod podzolics; (3) *humid-warm temperate and tropical zonal soils*, including reddish- and yellowish-brown lateritics, red loams, 'latosols', terra rossa, krasnozem, and perhaps red-yellow podzolics; (4) *soils of the forest-grassland transition*, with degraded chernozem, non-calcic brown and Shantung brown soils; (5) *the dark coloured soils of the subhumid grasslands*, which include reddish and normal prairie soils (brunizem), chernozem, chestnut and reddish chestnut soils; finally (6) *light coloured soils of arid areas*, including reddish brown, brown, grey (serozem) and other desert soils.

The zonal concept has had many precursors. It is frequently confused with Marbut's classification and often accredited to Dokuchayev, to whose scheme it bears more than a passing resemblance, for it is derived from it. Dokuchayev formulated two schemes, one in 1886,[26] the second in 1900.[27] He had three *classes* of soil—normal, transitional and abnormal. Of *normal* soils there were seven *types*, each with its own 'zone' (in parenthesis)—tundra dark brown soils (boreal); light grey podzolized (taiga); dark grey soils (forest steppe); chernozem (steppe); chestnut (desert-steppe); ærial and grey soils (desert); and red soils or laterite (subtropics and tropics). His *transitional class* had three types—dry moor and meadow soils; rendzinas and alkali soils. There were also three types of *abnormal soils*—bog, alluvial and æolian soils.

It was Sibirtsev[28] who introduced the terms 'zonal', 'intrazonal' and the concept of 'azonal' soils; as well as 'rough' and 'coarse' soils now called lithosol and regosol. He omitted peats. The term 'normal' has, of course, been dropped, for all soils are normal to their causative factors.

In all these early schemes 'type' corresponds to suborder or to group, and they are named almost solely by colour rather than by reference to genetic factors. The present Russian categorization[29-32] is a genetic system somewhat at variance with the American. *Genetic soil type* is equivalent to group. *Subtype, kind or genera* corresponds to family, and are distinguished according to texture. Then come *species*, distinguished by intensity of process—eg depth of humus penetration or degree of podzolization or salinization. Lower categories are *varieties* and *ranks*.

Later (post-1930) schemes of soil classification[33] sought to incorporate some reference to the weathering regime or balance of weathering processes. Given mountain, alluvial and tundra soils as special cases and gleization as an intrazonal process, there are four such regimes: (1) The temperate zone of *siallitic* weathering, confined to rocks with aluminosilicates, somewhat desilicified, and with illite dominance. This view excluded soils on sandstones as intrazonal 'psammitic' soils.[34] (2) Limey substrates had *calcitic* soils, comparable to pedocals, but including rendzinas. (3) Tropical *ferrisiallitic and ferric* soils, including the red soils of hot deserts, the red tropical loams of savanna areas and terrarossa. (4) *Ferrallitic* soils of moist tropical and equatorial regions, intensely hydrated, with no reserve of weatherable minerals; aluminosilicates disintegrating to provide insoluble Fe and Al oxides.

This approach is a dynamic one, reflecting weathering processes as much as the profile formed. Later additions are halomorphic, allitic and acid-humus soils, as well as 'raw mineral soils'. A considerable refinement and advance on such a scheme is the one proposed by Duchaufour and Aubert[35] with ten morphologic divisions, which are of equal rank and related to present profile morphology, rather than having an assumed place in some genetic sequence (see also Duchaufour, 1960, p. 206–7).

I now return to the *great soil group*, defined as 'a group of soils having a wide distribution and a number of common fundamental internal characteristics'. Groups have multiplied at an alarming rate, for while Dokuchayev and Sibirtsev (1900–1) had twelve groups for the whole world, now, for eastern Europe alone, there are sixty groups and many subgroups.[36] The USA has thirty-six groups, with provision for more in the tropical areas of Puerto Rico and Hawaii and in the tundra of Alaska.[2,37]

Doubt exists as to the status of some soil groups, especially the brown forest, the black cotton and the terra rossa soils which are considered by some to be calcimorphic intrazonal soils, and by others as zonal soils. Similarly, though group names are descriptive of colour and sometimes evocative of process, they are inconsistent. Some refer to the colour of surface horizons, some to that of the B and C horizons, some to the associated vegetation (prairie soil) or landscape (tundra soil). The terms imply prior knowledge of the soil in question and do not describe its total profile. Other objections are that the formative process is assumed, rather than implicit; and that shallow or mountain forms are not distinguished from deep, well developed forms.

Finally, though zonal soils are related to latitude, no attention is paid to longitudinal contrasts of western maritime, central continental and eastern monsoonal climates; nor to the transitional latitudinal zones between tropical and temperate, and between temperate and arctic, zones. Hence the zonal concept is geographically unsound.

To overcome the ambiguities and deficiencies of zonal terminology, and to pay more attention to the role of parent material, Kubiena[38] formulated a systematic classification of the total profiles of forty global *types* occurring in Europe. Each type (=group) is divisible into subtypes and then into *varieties*. The types are assembled into *classes* and then further into one of three *divisions*— (A) sub-aqueous, (B) semi-terrestrial and (C) terrestrial soils. His system of classification is a natural one, ordered by all soil properties, unlike previous systems which are artificial, for they classify soils according to properties selected as significant at the outset.

Division A has two classes—(AA) subaqueous soils not forming peat, and (AB) fen peat.

Division B has six classes of 'flooding or groundwater' soils: (BA) raw soils; (BB) anmoor and marsh; (BC) high moor peats; (BD) salt soils; (BE) gleys and (BF) warp soils; alluvial soils, derived by erosion of 'terrestrial' soils, *borovina* (rendsina warp), *smonitza* (chernozem warp) and vega soils.

The third division, 'C'—the better-known terrestrial soils—has ten classes: (CA) climax *raw soils*—with three types: *rawmark* (developed arctic and mountain soils); *yerma* (dry deserts); and *syrosem* (in temperate areas on newly exposed parent rocks); (CB) *rankers* with AC profiles on lime-deficient rocks; (CC) *rendsinas* on calcareous rocks; (CD) steppe soils with four types—*serosem*, *burosem* (grey and brown sub-arid soils), *kastanosem* and *chernosem*; (CE) terra calxis or 'altered soils on calcareous rocks', with two main types—*terra fusca* and *terra rossa*. (CF) plastosols—*the red and brown loams* of southern Europe; (CG) latosols; (CH) brown earths; (CI) pseudogleys (soils on clays with aerated brown A horizons and mottled subsoils), and (CJ) podsols—greatly subdivided, with twenty different sub-categories.

The nomenclature is of the simplest; each *type* has a name, a *noun* which 'may come from any language, may have any meaning, its only function is to be a proper name' and be as short as possible. The names of *subtypes* are formed by adding an adjective or prefix—

degraded chernosem, cryptopodsol, or else a genitive noun—*gley-podsol.* The varieties (which roughly correspond to series) are characterized by further additions and the order of words in the soil name thus becomes—variety, subtype and type, as in *mull-like xerorendsina* or *eutrophic calc braunerde.*

A feature of the scheme is that it uses the diagnostic horizon sequence of the total profile. One has ABC soils as fully developed eluvial-illuvial terrestrial soils with a basal weathering complex, or (A)C or (A)G profiles of soils in which organisms are invading terrestrial or semi-terrestrial sites. Other combinations are AC soils (with no illuviation), and A(B)C soils with weak, possibly mechanical, illuviation. Soils with surface crusts have B/ABC profiles; AC/ABC soils are polygenetic. From this one may develop diagnostic profile formulae for either Kubiena's types or the zonal groups—for example, LFH, A2, Bh, Bs, B2, C for a humo-ferric podzol; Ah1, Ah2, B, C for a brunizem; and LFH, Ahg, Bg, B2, C for a surface water gley. (See Fig. 5.)

Many of the deficiencies of the zonal scheme have also been met in the Seventh Approximation (7A),[2] which classifies all soils solely on those internal properties which either affect soil genesis, or result from it. The ten orders in its highest category are as rigorously defined as the lowest series (though this is true of Kubiena's scheme) and far more rigorously than the mere conjunction of a geographic name with a textural class. Similarly, the terms used (*typipsammentic halumbrept* or *natrustalfic mazustert*) describe the morphology exactly (once they have been learned), as well as connoting the quantitative dominance of the characteristic mentioned.

The 7A includes cultivated soils as equal members of the classification on the ground that the significance of changes of texture of the Ap layer is pragmatical. Other features are the disappearance of intrazonal soils and their reallotment to the surrounding natural orders as *aquic, andic* or *rendic* suborders or *halic* groups; and the characterization of many tundra soils at group level.

The Seventh Approximation has ten *orders* of soil:

ENTISOLS   *recent* (=azonal) soils; some humic gleys; rankers.
VERTISOLS   mixed or *inverted* soils = grumusols; margalitics.
INCEPTISOLS young soils on new surfaces; tundra; andosols.
ARIDISOLS   desert, sub-arid and associated saline soils.

MOLLISOLS   *soft*, subhumid calcimorphic soils; chernozem, bruni-zem, rendzina; brown forest and associated gleys and saline soils.

SPODOSOLS   wooded or *ashen* soils, mainly podzolized.

ALFISOLS   the other *pedalfers*, with mull, or else lessivé; grey-brown podzolic; degraded chernozem; planosol.

ULTISOLS   highly weathered ferruginous soils; red-yellow pod-zolics; red-brown lateritics and other hydromorphic variants of the *ultimate* stage of weathering.

OXISOLS   Ferrallitic soils—laterite and 'latosols' with free Al_2O_3, Fe_2O_3 and other *oxides* of Ti and Cr.

HISTOSOLS   Hydromorphic organic soils (bogs)—peaty *tissues*.

Suborders number 29, though oxisols and histosols are as yet undivided. Addition of one of fifteen suffix elements (eg *aqu*=wet; *ferr*=iron-rich; *orth*=common) to the root of the order name gives the name of the suborder as in *aquent* (waterlogged recent soil); *orthod* (podzol) or *boroll* (northern chernozem). The suffixes evoke some primary physical or chemical soil property which influences its drainage state, some textural singularity, or a climatic or biotic influence—for these control the rate or direction of soil development.

The *great groups* are named by adding a further short suffix to the name of the suborder. For the eight orders studied in the USA there are 105 great groups. There are forty-one mnemonic elements for group names and more will appear, applicable to tropical soils. The most used are *agr* (cultivated), *brun*, *calc*, *cry* (frozen), *dur* (hard), *eutr* (-ophic), *hal* (salty), *norm-* and *verm-* (worm-mixed). The groups are then subdivided into families, according to texture, mineral form, reaction, density and by temperature range at 20 cm depth.

There are also diagnostic subsurface and surface horizons for soils. The six most common surface horizons or *epipedons*—'mollic', 'anthropic', 'umbric', 'histic', 'ochric' (light) and 'plaggen' (man made) epipedons. The six diagnostic subsurface horizons are 'argillic' (illuviated clay), 'agric' (plough pan), 'natric', 'spodic' (iron-enriched), 'cambic' ((B))—in its limited sense of the loamier horizon of a developing brown earth—and 'oxic' (an horizon weathered to sands or to 1:1 clays). Various diagnostic pans are also recognized—eg, 'calcic', 'gypsic' and 'duripans'.

This system has had great impact on soil science.[39] Though at first sight learning Russian might seem a pleasant and relevant alternative to grappling with its nympholeptic nomenclature, the simplicity and directness of the 7A become apparent with usage.

1 B. E. Butler, Diversity of Concepts about Soils, J. Aust. Inst. Agric. Sci., 24, 1958, 14–20

2 *Soil Classification, A Comprehensive System, Seventh Approximation.* USDA, 1960

3 I. P. Gerassimov and E. N. Ivanova, S and F, XXII, 4, 1959, 239–48. Also G. A. Hills, *The Classification and Evaluation of Site for Forestry*, Ontario Dept. Lands Res. Rept., 24, 1952

4 *Soil Survey Manual*, USDA Handbook, 18, 1951

5 J. Thorp and G. D. Smith, SS, 67, 1949, 117–26

6 *Soil Survey for Land Development*, FAO Agric. Ser. 20, 1953

7 R. L. Pendleton and D. Nickerson, SS, 71, 1951, 35–43

8 A. J. Taylor, Geog., 45, 1960, 52–67

9 K. D. Glinka, *The Great Soil Groups* . . ., 1937, p. 70

10 A. Muir, *The Podzol and Podzolic Soils*, Adv. Agron. 13, 1961, 1–56

11 S. A. Wilde, *Forest Soils*, Ronald, NY, 1958, p. 210

12 G. W. Robinson, JSS, 1, 1949, 50–62

13 K. K. Gedroits, *Kolloidchemie*, 1929, 1–112

14 G. W. Robinson, *Soils, their Origin, Constitution and Classification*, Murby, 1949, see p. 408

15–16 C. F. Marbut, Soils of the United States, in *Atlas of American Agriculture*, Part III, 1935

17 V. R. Voloboyev, *Pochvy i Klimat*, Izd. Akad. Nauk. Azerb. SSR, Baku, 1958

18 H. Jenny, *Factors of Soil Formation*, McGraw-Hill, 1941

19 G. R. Clarke, *Study of the Soil in the Field*, OUP, 1957

20 *Soil Survey Method*, NZ Soil Survey Bull. 25, 1962

21 R. W. Simonson et al., TICSS, 7, 196C, V. 18, 127–31

22 F. F. Riecken and G. D. Smith, SS, 67, 1949, 107–15

23 B. W. Avery, A Classification of British Soils, TICSS, 6, 1956, V. 45, 279–85

24 M. Baldwin et al., *Soil Classification*, in USDA Yrbk., 1938, 979–1001

25 C. B. Wells, J. Aust. Inst. Agric. Sci., 26, 1960, 290–1

26 V. V. Dokuchayev, 1886, 'Materials for the Evaluation of the Lands of the Nizhny-Novgorod Government. (Collected Works, vol. 4, Acad. Sci. Moscow, 1950)

27 V. V. Dokuchayev, *Pektsie o Pochvovedenie*, 1900, in *Collected Works*, vol. 7, 257–96, Moscow, 1955

28 N. M. Sibirtsev, *Pochvovedenie*, St Petersburg, 1899, in *Selected Writings*, vol. 1, Moscow, 1951

29 I. P. Gerassimov, Die Erde, 1, 1963, 37–47
30 J. J. Basinski, JSS, 10, 1959, 14–26
31 E. A. Ivanova and N. N. Rozov, TICSS, 7, 1960, V. 11, 77–87
32 I. P. Gerassimov, in *Soviet Geography*, AGS, Spec. Pub. 1, 1962, 111–17, and see ref. 3, this chapter
33 E. H. del Villar, *Soils of the Lusitano-Iberian Peninsula*, Murby, 1937
34 A view still held by many Iberian workers
35 G. Aubert and P. Duchaufour, TICSS, 6, 1956, V. 97, 597–604
36 I. V. Tiurun et al., TICSS, 7, 1960, V. 5, 36–43
37 M. G. Cline, SS, 67, 1949, 81–91
38 W. L. Kubiena, *Soils of Europe*, Murby, 1953
39 P. Duchaufour, Soil Classification, JSS, 14, 1, 1963, 149–55; J. W. Muir, JSS, 13, 1962, 22–30. See also the special issue of Soil Science, 96, 1, 1963, 1–67, and I. P. Gerassimov, Sov. SS, 6, 1962, 601–9

10

Azonal soils or entisols

THE morphology of azonal soils is fairly uniform and they are usually classified according to the origin of their parent materials rather than by the processes of soil formation, which have hardly begun to act.

All three azonal groups—lithosols, regosols and alluvial soils—lack B horizons, and shallow A horizons are darker than C horizons because of additions of organic matter. *Regosols* are defined as thin soils on unconsolidated materials; *lithosols* as thin stony soils, developed on rock with (A)C profiles. With further development, *litholic* AC soils form. 'Alluvial soils' are those of active floodplains which receive additions of alluvium on flooding.

Lithosols usually occur on slopes with excessive, often erosive, runoff; little water entering the soil to promote leaching and weathering. Sites are dry, even in perhumid climates, and little plant debris is provided or retained. Surface horizons are usually grey—darker if organic matter is retained, lighter and coarse textured if surface wash is intense. The colour of the C horizon closely reflects that of the parent rock. Most lithosols are coarse calibred sand and fine gravel, and contain little fine matter. Depth of soil depends on how much recent erosion has taken place and on the weatherability of the parent rock. Man-made lithosols occur on eroded land, with (AC)C profiles, unsuited to further intensive cultivation, being shallow and drought prone.

Rankers[1] are thin lithosols on acidic rocks—humus-silicate soils. *Proto-rankers* have thin slightly decomposed micro-organic residues on quartz. *Tangel-rankers* develop in very acid conditions, as under broom or juniper litter in the alpine belt.[1] A distinction is made between *mulliform-rankers* if the organic debris is quite separate from the mineral substrate, while *mull-rankers* occur if humus and fine mineral debris are mixed in an A, AC, C profile. A lithosolic *grey ranker* develops into a litholic *brown mull-ranker*, then to *podzolic brown earth* and finally to *podzol* as an example of the sequence from an azonal to a zonal soil on siliceous rocks. Likewise a *thin rendzina* on a steep limestone scarp is an azonal soil developed from a protorendzina and will form a *degraded* (*or brown*) *rendzina* if the slope is stabilized. With time a *calcimorphic brown earth* may develop, with an A(B)C profile.

Soils on tephra and other soft, acidic, volcanic deposits such as rhyolite tuff, quickly acquire a zonal character; those on basic volcanic materials acquire one much more slowly. The initial black soils on basic volcanic material are termed *andosols*[2] and are associated with *grey hydromorphic soils*[3] in waterlogged areas. In Java (8°S)[3] *andosols* occur on the most recent deposits at considerable altitude, and alter rapidly with age (especially on lower slopes) by addition of organic matter. The dark brown A horizons deepen, developing into brown, red-brown, and finally to red tropical soils. In Nicaragua (13°N) andosols change to grumusols;[4] in Japan (38°N) to brown forest soils and then to podzols.[2]

The Ah horizons of andosols have low bulk density (0·3), abundant plant nutrients, and many earthworms, and are rich in organic matter with 8% on average, attaining 30% in the darkest soils.[3] Mineral A2 horizons are silty, water holding and hydrated; and clay migration forms a panned (B) horizon. Clearly andosols can be considered as lithosols, or regosols, or as intrazonal soils, and quickly assume a zonal character—except perhaps in Iceland,[5] where they are kept 'fresh' by frost action and additions of wind-blown material.

Regosols consist of deep, soft mineral matter in which only A and C horizons are developed, and are confined mainly to sand dunes, loess and recent glacial till. *Soils on dunes* are thin and grey, consequent on rapid decalcification in humid climates. Only organic (LFH) and A layers change with time. B horizons do not develop and Ab horizons recur to depth.[6] Dunes are formed in many zones, zonal in hot deserts, possibly intrazonal forms in arctic and in temperate areas.

Loess has a silty texture and largely consists of light minerals such as carbonates, mica, felspars, and fine quartz. Median particle size varies, being finer and more rounded with increasing distance from the source. Peoria loess in Illinois has a median ø of 0·05–0·02 mm.[7] Loess in north Belgium near Ekloo has a median ø of 130µ, 55 km to the south at Waterloo it is 30µ,[8] but in the Condroz, a further 50 km south, it is 2–20µ.[9] A maximum size for loess is c 0·1 mm in, for example, the Dutch *dekzand* of niveo-aeolian origin.[10]

Loess is widespread in central Europe, north-central USA, southern USSR, northern China and Argentina. The most common azonal soil on loess has a decalcified A horizon of pH 6, with wider quartz/Ca ratios than in the C horizon. Soils on loess are well aerated, rapidly assuming zonality as grey-brown podzolics, brunizem or chernozem. Many loess deposits contain buried soils, a noteworthy example being sited south of Constanta, Rumania.[10b]

Soil development on till is slower than on loess due to freshness of minerals, higher contents of unweathered rock and compaction by ice. As a parent material *till* may vary greatly in stone and lime content, texture, compaction and mineral content. Hence one needs to distinguish between *source materials* of till which become mixed to form the *parent material* of soil. Some tills in East Anglia and eastern Sjaelland are lime-rich, the glaciers having passed over limestone tracts, while other tills in Northamptonshire or eastern Jutland are rich in acidic clay derived from Jurassic or Tertiary source materials.

Tills formed during the advancing stages of active ice sheets are compacted, well mixed and fine grained; till deposited by dead ice is coarser, ill sorted, less compact and has hummocky relief, with hydromorphic soils in hollows and over-drained soils on hummocks —the same arrangement is found in terminal moraine country. Till in areas of retreat and final ablation is stony and of local origin, as in northern Britain and north Sweden.

Present soils on till, related to formerly glacierized areas, are mainly leached soils of temperate or cold areas, though chernozem and chestnut soils are known on till in north-central USA.

Alluvial soils may be classified by texture and by textural profile; or by zonality or degree of evolution. They are usually light-coloured *entisols* on level sites. Reduction processes dominate in the subsoil, though only slightly so, for the water in floodplains is rarely stagnant. Alluvial soils of fine texture or with very shallow water tables have low vertical permeability and have perennially moist surface layers rich in organic matter and reduced subsoils. If coarse textured, the soil is better drained and surface layers dry out more readily. Many, but not all, alluvial soils are rich in lime (Mississippi, pH 6–7·5) derived from leaching and erosion of terrestrial soils.

Alluvial soils have multi-phase profiles, with alternating layers of gravel, fine material and organic debris. These 'horizons' are rarely continuous away from the surface and the material varies in origin as well as texture. In large drainage basins with broad floodplains alluvium is far-travelled, greatly mixed and water-worn. In the narrower alluvial tracts of low-order streams it consists of material very similar to adjacent soils. Thus one may refer to 'general' and 'local' alluvial soils of varying age.

A factor in the differentiation of alluvial soils is their position in relation to the main stream and to the adjacent higher or 'stable' land. Five principal site areas may be distinguished, duplicated on each side of the river[11]: (1) *levées* of coarse debris with a deep water table; (2) *sandy alluvium* with a water table at moderate depth in the floodplain centres and well mixed plant and mineral debris as a granular mull; (3) the moist strips close to levées; (4) sites nearer to the side slopes than (2) with fine, dry, often ochreish material; and (5), sited nearest the valley sides, moist and peaty soil, the alluvium thin and fine. Slope-wash from the stable land is added as well as spring water. Area 5 has 'boggy, pre-terrace soils',[11] *nassgley*[12] or 'sod gleys' with an AG profile. Peat often forms in area 5, and is more extensive in the large floodplains of colder climatic zones. Such *bog peats* also form in depressions and oxbows, with structureless silty clay AG profiles. Vigorous plant growth causes deep F2 and H or else T horizons.

Changes of river course are the main causes of erosion of alluvial soils, and abandoned stream channels provide coarse parent materials.[13] Annual flooding of alluvial straths gives deposition of fine material; infrequent intensive flooding gives deeper layers of coarser debris and a multiplicity of buried profiles. Augering often reveals (F), C, Ab and Cb profiles with the new material at the surface.

Alluvial soils evolve toward a zonal soil by decrease of hydro-morphism or by intensification of zonal conditions. As soil water regimes are governed by rainfall as well as by ground water, then, if the water table is lowered there is enhanced leaching by rain and the zonal influence is felt more keenly. A horizons deepen and vestiges of illuvial (B) develop if leached materials do not go directly to the river in solution. Decreased hydromorphism is also achieved by reduction of river discharge and the shrinkage of streams.[14] Incision forms terraces and a narrower floodplain; shrink-age gives a narrower floodplain and drying at the same morphologic level.

Soils on terraces are older, better drained and show more zonal tendencies than floodplain soils. Terrace edges have the deepest soils, their backs are less well drained. Many terraces in Europe are cultivated and have dark grey sandy loam Ap, A2, B(t), B2, and C profiles.

Yet alluvial or terrace soils existing in one zone may have been transported from another weathering zone. The alluvial plains of the subtropical Mekong and Ganges, derived from mountainous or arid areas can be richer in bases than surrounding terrestrial soils.

A distinctive alluvial soil is the *gyttja* of north Sweden and Finland,[15] consisting of organic and alluvial or lacustrine layers. It has a high S content and high porosity, but low permeability due to underlying Ancylus clays. Though gyttja soils are rich in nutrients they remain in pasture for they have little lime and much oxidized S at the surface. They have very low pH (2·5) and are toxic to deep-rooted plants. At depth, reduced iron sulphides or free S are present in a black layer. The soils illustrate phase I of Polynov's scheme in a cool temperate area.

Vega[1] is a term variously applied to red or brown alluvial meadow (warp) soils with groundwater at *c* 2 m. They are irrigated for pasture in north German valleys such as the Ems,[12] for primeur crops in the huertas of eastern Spain.[1] Fe content is high (2%–4%). organic content is 1% and pH varies. They are also called *acid brown auenboden* in Germany, but are red *kalkvega* in Spain, of higher pH. In central Europe *vega* are associated with gleys with groundwater at 20–100 cm, which are richer in organic matter (3%–8%), have higher pH and more Fe and Mn. Adjacent *saturated anmoorglei* (humic gley) have 15%–30% organic matter in an Agh horizon, and high C/N ratios (17·5 to 30).

Marsh soils form a sub-group of alluvial soils. In Holland, classified according to origin, age, lime content or texture; there are four families of marsh soils[16]: (1) *river marsh*, the alluvia of rivers close to the sea, with low salt content (0·1%) derived from the river's solution load, which are often (2) *peaty*, and have silty clay texture; (3) *brackish marsh soils* (knickmarsh)[17] which have little or no lime at the surface and salt contents of 0·2–2·0%. They are old marsh soils at some distance from the coast, reclaimed in medieval times; (4) *marine salt marsh soils* on *schlick*,[17] or in new polders[10,17]; they are calcareous (5–15% $CaCO_3$) and saline (>2% salt) with pH *c* 8 (Fig. 5, 30–1).

Mangrove swamp soils and tropical marshlands on coasts (*poto-poto*) are a distinct intertropical regosol. A shallow layer of soft silt is trapped by roots on top of a black muddy layer *c* 1 m deep and with *c* 20% organic content.[18] Fresh mangrove swamp soils have neutral to high pH, and become acid by surface oxidation in the dry season or on drainage, acquiring a grey-brown colour. As in gyttja, reduction of sulphates at depth gives intensely black subsoils. Other tropical regosols occur on lacustrine and coastal dune deposits—the latter with ultrapodzols in humid areas—sandy soils on coral rag, and soils on pumice.

Azonal soils are many and varied. It is hard to draw the line between those with little or no genetic horizons and those with deep but vaguely differentiated horizons and, too, those of mountain and tundra lands related to similar regolith in which horizons may be well developed but only a fraction of an inch thick.

Soils of mountain areas are usually litholic, or else regosolic with parent materials respectively derived by *in situ* physical weathering or by mass movement and colluviation. Other parent materials are valley infillings of varying age, alluvium, and fresh glacial and peri-glacial deposits. Except for sandstones, shales and volcanic mat-erials, available rocks are hard and resistant to weathering. Soils of mountain areas are therefore shallow, have high stone and gravel contents, and are drought-prone as they have rapid absorption and transmission of moisture but little adsorptive power. Biological activity is slight; plant debris is sparse and it decays slowly. It

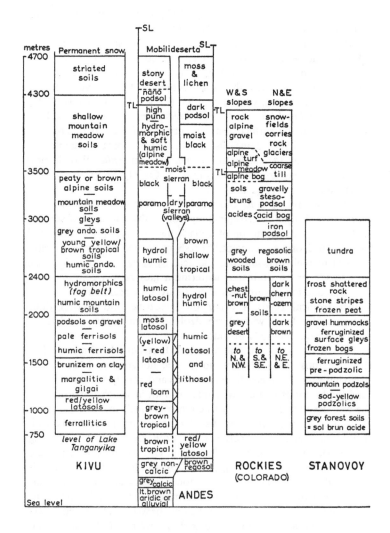

Fig. 6   Vertical zonality of mountain soils
SL—snow line; TL—tree line

usually remains unmixed as a mor or moder horizon, so *mountain turf* or *alpine meadow soils* form above the timber line.[19,20] *Turf* soils are better drained, are found on steeper slopes, and have a herbaceous cover. They have gravelly y/b, acid, Ah, A2, B1, B2, C profiles, with A+B extending to 50 cm, the C horizon a further 30–40 cm as fractured rock. If the turf is broken by erosion, barren *gravelly fields* result.

Alpine meadow soils occur at lower altitude on alluvial or colluvial materials. Typical sites are the perennially moist depressions or wide benches ('alps') which have willow or sedge vegetation. The soils have g/b, loamy, A, AC, Cg profiles. The A horizon is dark grey and thin; the AC is grey and 20 cm thick; the Cg is >1 m deep, is frozen, and has pH 5 to 6—high for most mountain soils away from limestone. With time such soils become peaty if they are not buried by creep or landslide material.

Mountain soils differ primarily because of variations of site, exposure and denudation balance rather than variations of parent rock, and zonal development occurs hypsometrically. Humus and hydromorphism increase with altitude, especially on benches. Zonal altitudinal sequences are given (Fig. 6) for humid-tropical,[21] dry-tropical,[22] subhumid warm temperate[20] and continental boreal[23] mountain ranges.

1 W. L. Kubiena, *Soils of Europe*, 1953, p. 163–77

2 I. Kanno, TICSS, 6, 1956, V, 105–9. The name is derived from '*An-Katsu-Shoku-do*', meaning 'dark brown soil'

3 K. H. Tan and J. Van Schuylenborgh, NJAS, 7, 1, 1959, 1–21 and 9, 1, 1961, 41–54

4 B. W. Taylor, Estud. Ecol. Nicaragua, 1, 1959, pp. 310–11

5 B. Jóhannesson, *Soils of Iceland*, Dep. Agric. Rept., B. 13, 1960, also B. Helgasson, JSS, 14, 1, 1963, 64–72

6 B. C. Barratt, Nature, 196, 1962, 835–6

7 G. D. Smith, *Illinois Loess, A Pedologic Interpretation*, Univ. Ill. Agric. Expt. Stn., Bull. 490, 1942, 139–84

8 Data after Belgian Soil Survey publications

9 R. Maréchal, Pédologie, Mem. 1, Gent, 1958

10 C. H. Edelman, *Soils of the Netherlands*, N. Holland Pub., 1950, p. 17

10b A. Conea et al., R.P.R. Stud. Tehn., Ser. C, 12, 1964, 11–44

11 Z. N. Gorbunova, Sov. SS, 1, 1961, 48–54 and L. I. Korableva, Sov. SS, 4, 1961, 387–94

12 E. Mückenhausen, 1960, profiles 54, (nassgley), 48 (braun kalk-vega), 49 (auenboden) and W. Baden, *Festschrift zum 75-jährigen Bestehen der Moor-Versuchstation in Bremen*, 1952

13 B. E. Butler, A Theory of Prior Streams . . ., Aust. Jnl. Agric. Res. 1, 1950, 231–52

14 G. H. Dury, Geog. Rev. 50, 2, 1960, 219–42

15 L. Wiklander and G. Hallgren, Ann. Ryl. Agr. Coll., Sweden, 16, 1949, 118–27 and 17, 24–36

16 W. Müller, Geol. Jb. 76, 1959, 11–24

17 E. Mückenhausen, 1960, profile 58 (knickmarsch), 57 (altmarsch), 65 (jungmarsch)

18 M. G. R. Hart, Plant and Soil, 11, 1959, 215–36

19 J. L. Retzer, JSS, 7, 1, 1956, 22–32 and 20. *Soil Survey of the Fraser Alpine Area*; Colorado, USDA, Ser. 1956, no. 20, 1962

21 The profile for Kivu is based on: I. A. Denisov, Sov. SS, 6, 1961, 604–8 and on A. Pecrot, Pédologie, IX, 1959, 227–37; A. van Wambeke and L. de Leenheer, Med. Landbohogesch. Gent, 26, 1962, 697–812; also Pédologie, XI, 2, 1961, 289–352

22 F. Quevedo et al., TICSS, 7, 1960, V. 13, 97–104; also E. Miller and N. T. Coleman, PSSSA, 16, 3, 1952, 239–44

23 V. M. Fridland, Poch., 11, 1959, 8–18; (Baikal); G. Haase, ZPDB, 102, 2, 1963, 113–27, (Mongolian P.R.); T. A. Rode and I. A. Sokolov, Sov. SS, 4, 1960, 384–91

The 'classical' work on mountain soils is H. Jenny, Hochgebirgsboden, in *Handbuch der Bodenlehre*, 1930, 96–118. A useful resumé is by A. B. Costin, JSS, 6, 1955, 35–50

11

Intrazonal soils

INTRAZONAL SOILS comprise three suborders—hydromorphic halomorphic and calcimorphic. In addition there are other soils of a world-wide distribution—anthropomorphic soils, soils on organic materials, and certain distinct horizons developed in special circumstances in zonal soil profiles which may be regarded as intrazonal.

Many intrazonal soils are sufficiently distinct in the appearance of their profile and in their use by man to be considered as separate groups within their suborder, though geographers must also approve the method of the 7A, in which they are linked with the adjoining zonal soils as aquic or halic variants. Gleyed soils may also be associated with the drier soils on adjacent slopes. In many cases gleyzation and salinization are processes which act on previously existing groups, rather than distinctive soils, except in extreme cases.

The most widespread intrazonal soils in humid climatic areas are hydromorphic soils—*gleys*—occupying lowlands, with slow runoff. They are found in tundra, temperate and moist tropical areas in depressions, gain runoff from adjacent uplands, in addition to rainfall, and have high water tables for all or part of the year. The subsoils of many halomorphic soils in depressions also show gley phenomena.

A gleyed soil is one with part or most of its profile waterlogged, therefore undergoing reduction instead of oxidation. Ferric oxides are reduced to ferrous salts, giving the soil a uniform grey or blue colour. Waterlogging may occur (1) at depth, or (2) near the surface, or, in some cases, (3) in the central part of the profile. In the first case a basal G or CG horizon forms. However, should the water level vary seasonally by more than 1 m a partially gleyed horizon with much mottling and some structural forms will occur above the true gley—G—as a Bg or in some classifications[1] a 'g' horizon. The mottling is caused by oxidation and reprecipitation of ferric iron along the most aerated parts or channels in the periodically waterlogged and mainly grey anaerobic layer.

Gleying also occurs in soils based on impermeable clay parent materials. Here rain or drainage water or both pass through the material only slowly or not at all. The upper part of the soil is therefore waterlogged at the season of greatest water supply. Impeded removal of surface water gives *surface water gleys* with organic moder-like surface layers and gleyed Ag and Bg horizons, but with C horizons appropriate to the zonal group or to the colour of the parent materials (Fig. 7, profile 2). These have sometimes been termed '*stagnogley*' soils, if subject to prolonged wetting.

In contrast, superficially effective surface drainage but impeded subsoil drainage produces *groundwater gleys* (Fig. 7, profile 1), with thin oxidized A horizons and little or no organic debris; and reduced and compact subsurface grey blue G horizons and perhaps an intervening mottled Bg horizon. These have been termed *typical gleys*, of variable but usually neutral to slightly acid reaction, and a massive or blocky structure.

The name 'typical gley' is taken to point the contrast with pseudogley soils—those with a temporary perched water table above a compact horizon. These are, of course, slightly degraded zonal soils, of a brown earth or sometimes podzolic nature, subjected to temporary hydromorphism during their development.

Unfortunately the terminology of gleys is confused, as in the case of podzols, by using the same word for a process and for the many different profiles in which the process can act to greater or lesser degree. Indeed the concept of gleying held by many English surveyors is one with the two main terms changed around; a 'ground water gley' being the one moistened or even flooded at the surface by rise of *groundwater* in valleys, while the 'surface water gley' is the ABgG profile, in which water content increases in the subsoil by seasonal additions of surface (ie rain) water, though the actual surface may never be gleyed. If one recognizes the subtlety, but not the logic or semantic adequacy of this latter view, and reverts to the former interpretation, a general horizon formula for a *groundwater or typical gley* would be: (H), Ah, A2 (or Ap), B(g),

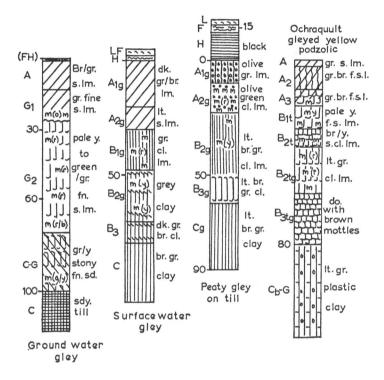

Fig. 7 Hydromorphic soil profiles, from Scotland and Georgia, USA. The organic-mineral interface is at the same level. Depths are in cm. Colours and textures are given, and structures are represented diagrammatically

B2g, G1-2, CG, with reduction and gleying increasing with depth— while the profile of a *surface water gley* (*pseudogley*) would show a diminution of gleying with depth, the gleying being due to water added at the surface and causing reduction there. Unlike the typical gleys, they are not perpetually anaerobic at depth. The profile formula is therefore: LFH (t), Ahg, A2g, B2(g), B3, C.

Temperate gleys may also be divided into calc- (Fig. 5, 17) and non-calc-gleys, and in North America[2] they are characterized according to progressive hydromorphism into the sequence:

(1) *Low humic gleys*, which are very acid, with thin mor and markedly eluviated A and illuviated B horizons with (H), Ah, BG, C profiles;

(2) *Humic gleys* (Fig. 5, 29), which are very poorly drained soils, virtually permanently waterlogged and anaerobic, developed under swamp woodland or herbaceous marsh. They have thick dark LFH and soliflual Ah layers, occasionally a thin A2 horizon and thick, dark, slightly acid to neutral G horizons;

(3) *Half-bog* (or *peaty*) *gleys*, which have deep H or T layers and are associated with peaty gley-podzols with thin iron pan;[3]

(4) *Bog gleys*, which occur if HT layers are thicker than 25 cm.

Other forms of gley are the tropical gleys, very similar to temperate gleys, but usually darker or black in colour and very mottled; while, in arid areas, the ground water is enriched with soluble salts and saline soils occur.[4]

Gleization is thus the process by which a blue, grey or olive coloured horizon is produced by reduction consequent on waterlogging at depth, or by excessive moisture at the surface.[5] It need not necessarily imply the presence of a mottled layer. The resulting G horizons of permanent reduction are sticky, structureless and compact when wet, but may dry out slightly to form blocky, mottled, rust-coloured G1 layers. Within deeper G1 layers Fe^{+++} is rapidly reduced, sulphates form sulphides and nitrates form ammonia. Reduction is due to low O_2-tension, a condition which often follows waterlogging, and micro-organisms and plant roots may also decrease the oxygen content of the water as fast as it is renewed at the surface.

Bloomfield,[6] in experiments mixing clay with water and various forms of plant debris showed that little Fe was dissolved in the presence of humus or peat, but that leaf and grass debris caused gleying, presumably by products of grass fermentation. Reduced iron is therefore more dominant under grass than under peat or raw humus, with Fe^3/Fe^2 ratios of 7 to 14 in humic gleys and 5 to 9 under meadow gleys.

The pale coloured A layers of some gleys are similar to the A2 layers of moist podzolized soils, though neutral rather than acid in reaction, for bases may not be completely removed, as in the podzol. Fe and Mn accumulate at the base of the dried A horizons in the gley if the annual cycle of wetting and drying is regular; yet they form concretions in a deep, ochreous, mottled layer if the annual variation of the level of waterlogging (and oxidation-reduction) is great, especially on better drained sites.

Finally there are many transitional zonal-hydromorphic soils, affected by subsurface water tables—groundwater podzols, groundwater laterites, gleyed brown earths, and groundwater rendzinas; some being more gleyed than others.

Organic soils follow on logically from gleys and also vary enormously. They intergrade with bog and with peaty gleys. Where peat occurs over large areas, as in the Canadian muskeg and in central Siberia, it has its own geomorphic forms and great depth (>2 m).[7,8] Climatic (strictly 'zonal') peats range from tropical to tundra forms, while local (aclimatic or azonal) peats occur in most zones as *basin peats*, wherever excess stagnant water accumulates. *Hill or moss* (*zonal*) *peats* form on sites having external drainage in areas of high rainfall and low evaporation. They have low bulk density (0·06) and are acid (pH 3–4).

Basin peats are subdivided into (1) *eutrophic fen* (=Niedermoor), with pH 5–8, associated with calcareous or base-rich water, and rich in ash; (2) *raised moss* (=Hochmoor), with pH <5, forming as a compact brown peat wherever it rises above groundwater level; while (3) is the *oligotrophic or acid low moor*, in which drainage water is base-deficient—pH <4·5. There are also many transitional forms between hill and basin peats.[9]

Soil formation on peat occurs on dehydration ('ripening'[10]), on drainage, or by plant root extraction of moisture. These reduce its mass and increase its permeability. Disintegration, biologic moulding and leaching ensue, and illuviated organic horizons form. Eventually stratified upper 'muck' or moder-like Aht horizons form, underlain by a compact, moist AT and an illuviated TBh at the base of the 'soil'. The substrate is the original T2–T3 sequence. LFH layers may also occur at the surface if plant growth is active.

Compaction of peat is often rapid. In the English Fens, Holme Post showed an average compaction of 3 in/yr between 1851 and 1892, now diminished to 1 in/yr.[11] In the Sacramento delta[12] the rate

E

is also c 3 in/yr, mainly the result of oxidation, burning and wind erosion rather than compaction and cultivation.

Tropical peats are a distinct hydromorphic type, very deficient in ash and with high organic colloid contents. All are acid—the *Machongo* of S. Mozambique (18°S) 10m deep and with pH 4·4; the *Rawa Lakbok* of S.E. Java (7°S), a 6 m deep topogenous peat with pH 3. The peats of the old delta of the Agneby (Ivory Coast, 5°N), have pH 2·7–3·5, C/N ratios of 20–50 and 75% organic matter (by weight). In coastal Guiana (7°N) the *Pegasse* peats (<40 cm thick) with pH <4, have Al toxicity. This peat is underlain by clay and tends to float when irrigated, so must be deep-ploughed before use.[13]

There are many distinctive forms of cultivated hydromorphic soil. *Paddy soils*[14] are similar to gleys, though flooding (at planting) and aeration (at harvest) occur several times a year in contrast to the annual wetting/drying cycle of most temperate gleys. The leaching of bases by flooding is balanced by fresh weathering in the dry periods. Paddy soils occur on many different parent materials, for example on coastal muds, or on alluvia derived from many sources, especially acid red and yellow tropical soil material.

In lower Burma (17°N), with 2600 mm rain, a 50 cm Ap layer overlies 2 m of G1 and 5 m of G2. The A layers are acid (pH c 5), with c 2% humus, exchangeable Al, and illite dominates. The basic G1 cracks and is oxidized in the dry season and shows weak structural forms and rust mottles. The G2 is bluish grey, with whitish siderite concretions ($FeCO_3$) if the soil is humus-rich.[15] Groundwater is at 3–5 m at the end of the dry season and G11, G12, G21 and G22 horizons may then be discerned, with varying colour and structure.

The profile of paddy soils is mainly textural. Surface textures vary with microrelief and with distance from the main river. A compact Ap1 is usually underlain by a granular porous Ap2 in long cultivated, well manured areas, underlain by a massive or blocky Ap3. The percentage of silt+clay of the A layers is never less than 55% and may reach 95% in the G2. The G horizons are poorly permeable (c 1 mm/day) and retain up to 80% of their field moisture capacity in dry periods. They are mainly basic (pH >7), and base saturation of 75%–90%.

Degraded paddy soils have lower humus, lime and base contents, while SiO_2 and R_2O_3 accumulate as concretions at 1 m depth. The Ap loses its fine texture and its capacity to shrink and swell; it hardens in the dry season, and is bleached and markedly acidic.

In upper Burma (22°N), with 800 mm Ø rain, *alkaline paddy soils* occur, with carbonates in the profile.

Plaggen soils[16a,17,18] are 'anthropomorphic soils' (existing entirely owing to the activity of man) and are thick, artificial layers of black or brown organic material burying sandy, often podzolized, substrates. Thus a profile sequence would be: Ap1, Ap2, (B)Ap, IIAh, IIAe, IIBh, IIBs, IIC. Such soils are common in north Germany and in Brabant, originating in the medieval practice of folding animals on layers of gathered turf or ling placed on harvested plots. The manure added to this debris gradually formed a cultivable layer, in time thickening to >1 m, thickest near to the villages, and far more fertile than the underlying soils. C-14 dating of these soils[16b] shows the base of the 'sod' to date from *c* AD 700.

Vitisols[19] are another anthropic soil, varying greatly in their seasonal appearance, with thick organic and refuse layers in the spring over stony CR horizons. Vitisols occur upon slate or rendzinic material, or upon subpodzolic yellow soils with reddish 'alios' or Bfe horizons in Germany and S.W. France respectively.

Intrazonal soils resulting from long lasting effects of parent material are many and varied. *Grumusols*[20] occur on clayey parent materials which crack on drying and slough at the surface on rewetting. This causes heaving of humps and ridges, the formation of micro-depressions and slickensiding at depth. *Gilgai* relief[20,21] thus forms on flats and gentle slopes. Profiles vary slightly on knolls and in depressions,[21] but are diffuse, and humus decreases gradually downwards due to mixing. The average soil depth is 2 m, with ACR profiles, the R the unmoved, unweathered, clay. They occur on dark grey or brown poorly drained calcic or non-calcic materials —for example on ultrabasic rocks and on former playa lakes—and are extensive in the Riverina of Australia. Grumusols, similar to tropical margalitic soils, have the same drawbacks to practical use

Calcisols form on highly calcareous rocks.[22] With time lime is leached away and A horizons thicken. Usually AC soils form, for the leached material is soluble and not readily redeposited, at least not as a B horizon. Calcisols are common as shallow greyish soils in sub-arid areas, on playa-side benches, with lime derived from groundwater or from seepage. In south-west USA such soils have Ap, AC, Cca profiles, the Cca layer being of pinkish *caliche* or lime pan with 25% $CaCO_3$, and replete with worm casts and insect burrows. Such soils are often under irrigation for cotton.

The parent material for calcisols is highly variable and its nature determines the colour, texture and profile of the soil far more than do external influences, except for the effect of slope angle on soil depth. Pure limestones in temperate and cold climates produce dark *rendzinic* soils, rich in organic matter, while cherty limestones have *brown calcimorphic soils* with A(B)C profiles. In subhumid or hot climates pure limestones give *dark reddish soils* with B horizons, cherty rocks give a yellowish-red soil, and sandy limestones a highly erodible yellowish-brown soil. In perhumid West New Guinea an 80 cm deep y/b silt loam has developed on Pliocene limestones (Fig. 13, profile 8).[23]

In many limestone areas waterlogging of soil in depression sites is counteracted by the permeability of the subsoil and the jointed nature of the rock. Soils in karstic basins, for example, may be podzolic if underground drainage operates, though, if impeded, as in the closed and alluviated basins of Thessaly, *calcareous-alkaline soils* form.[24]

Rendzinas develop in humid climates on parent materials with not less than 40% $CaCO_3$ and have dark humic or mollic epipedons. The primary stages in their formation have already been mentioned. By the time the A horizon attains 30 cm thickness, earthworms and microfauna are very active, forming clay-humus complexes in a crumb-structured A1ml, underlain by a granular A2.[25] All rendzinas show higher than neutral pH, with carbonates throughout the profile. Any sesquioxides are coagulated as soon as released, and no B horizons form. Yet degradation gradually forms a *brown rendzina* or a *brown calcimorphic soil*[26] with a (B) horizon. A *pararendzina* forms[27,28] if brown material is added by colluviation from upslope—for example from a cover of clay-with-flints—and ferric oxides colour the A2 and (B) horizons before bleaching occurs. Brown pararendzina also seem to be associated with sandier calcareous parent materials (Duchaufour, 1960, p. 238).

Brown carbonate-clay loams quickly form on limestone in warm sub-continental climates as in Hungary, the Balkans and the eastern Alps.[29] With clay formation and Fe incrustations they come to resemble the r/b clay loams of tropical areas. *Terra fusca* (Chapter 16) develop from these carbonate-rich materials and are the final stage of soil formation in south central Europe; but in the Mediterranean area terra fusca develop into *terra rossa*; or else a *rendzina* may develop into a *red clay loam* and then to *terra rossa*.[30]

Planosols are widespread on level plateau surfaces or broad gentle slopes on loess, till or wide alluvial terraces in central USA (Fig. 14, 7).[31] They are marked by deep clay hardpans or compact subsoils, have warm-season waterlogging, and water drains away very slowly both vertically and laterally. They have mollic epipedons and shallow, light-textured, bleached A2, for clay has moved downwards to compact the B horizon. Planosols have high base saturation and are dominantly grey, with a distinct Ap, Ah, A2, B1-B2t, B3(t), C horizonation, reaching to depths of 1·5 m. The A and C horizons are thin. Sheet erosion of the deflocculated A horizon occurs under intense rainfall on unprotected ground, even on 2° slopes, with the thick B1t then exposed as a 'gumbo' layer.[32]

Accretion gleys[33] are formed of fine clayey material gathered in shallow, waterlogged depressions on a till surface as a result of sheetwash of adjacent slopes. They are reduced soils, grey, with Ah, AG, G, D profiles.

A distinct form of soil horizon is the *fragipan*[34] which occurs in many temperate soils in formerly periglaciated areas, mainly within a layer of loamy material, especially in thin loess, or on soliflual till layers, or other material capable of forming cracks (Fig. 5, 14 and 16). Unlike planosolic horizons, fragipan layers are acid. They are sited in broad, well drained depressions or on erosion surfaces or spur remnants, and their characteristic feature is a thick Ct or Cx horizon (7A) enriched in clay skins or infillings of very fine sand and silt, often only along fine lateral and vertical cracks. The whole forms a brittle, platey or laminar-structured *fragipan*, with its upper boundary at 30–90 cm depth, extending to 2 m depth. The soils have high bulk density (1·7),[35] are low in organic matter and 'x' horizons are hard when dry, but soften and fracture on stress when wet. The grey infillings may outline fossil polygon features and have been considered to be fossil permafrost horizons,[34] or loess compacted by the weight of later ice. They are now recognized by many as genetic soil horizons developed by the translocation of fine material into cracks in the profile,[35] especially of grey orientated clay or silt into cracks formed by drying of subsoil layers (see 7A, Fig. 20).

Distinctive fragipan layers have been recognized in the Ardennes and in north-east USA within sol brun acide profiles, with profile formulas of Ap, B1, B2, C1x-C4x, C5 (7A, p. 94). Often the fragipans are so compact and impermeable that leaching or erosion of the upper layers is achieved by laterally moving water.

Saline and alkaline soils

In Kubiena's classification these soils are semi-terrestrial soils of former or presently-brackish water areas, and have a high salt content. Their type area is the land north of the present Caspian Sea, coincident with the former extent of that sea. They also occur in depressions in arid and semi-arid areas and are mainly dry at the surface in summer, though flooded in a wet season. Saline soils also occur in newly weathered materials in cool, even in arctic, continental areas. Most saline soils have slowly permeable substrates and saline groundwater at shallow depth, and are rich in some of the chlorides, sulphates or carbonates of Na, Mg and Ca.

There are three main groups of halomorphic soils: saline soils—*solonchaks*; the alkali soils such as *solonets*; and the leached or degraded *solod*. In addition there are transitional forms and many aridisols with distinct saline surface horizons. In all these soils salts are brought from groundwater by uprising capillary currents—from a maximum depth of 2·5 m in clays, only 0·7 m in sands.[4] If groundwater is at less than these depths saline crusts or superficial salt-enriched layers will form; if at greater depths, salt-rich layers form at depth in the profile, for the capillary currents are disrupted by drying.[4]

Halomorphic soils represent one of the most important soil resources of the world that remain to be used by man on a permanent basis. Of the drier areas of the world, 40% are subject to some form of *salinization*, with salts derived during the first phases of weathering or from wind- or rain-borne cyclic oceanic salt[4b]—as in

Israel, W. Australia and California. Halomorphic soils occur in sub-arid USA, in the Caspian—Aral basin (Kazakh SSR), in W. and S. Australia; in Hungary, Turkey, Sudan and N.E. Kenya, in W. Pakistan and western India, in parts of Spain,[38] the Dutch polders and other more local examples.

Usually more or less barren, but potentially highly-productive, most halomorphic soils are zonal soils which have been subjected to salinization. Regarding them as intrazonal is unsatisfactory, for salinization is largely limited to the transitional zone from sub-arid to arid climates. The more widely scattered local variants are inceptisols; intrazonal forms are located in depressions with saline groundwaters or on parent materials rich in salt, such as recent marine deposits.

Solonchak is a truly saline soil—'white alkali soil'—dominated by NaCl and NaSO$_4$ (>0·3% of the total soil). Chlorides and sulphates of K, Mg and Ca are also present, though Na forms >15% of the total exchangeable ions. If NaCO$_3$ or other carbonates dominate then *black alkali solonets* form (Fig. 8, 1–2).

Solonchak have high pH (>8), are poorly drained, and have highly saline groundwater at <2 m depth. On intense evaporation and capillary rise, salt deposition occurs at or near the surface and one, perhaps two, Sa layers form; such are toxic to plants, soft and friable when moist, and form a crust when dry. Adjacent shallow A layers are powdery when dry, have low density (0·8), and humus derived from Salicornia roots. The soil at depth shows massive gley-like features (Fig. 8, 1).[39]

Solonchaks originate in three ways:[37,40] (1) naturally as primary *soda solonchak* in depressions; (2) artificially—the result of irrigation by saline water without adequate drainage and groundwater control; (3) by the secondary though natural salinization of a pre-existing zonal soil. Kubiena cites other European forms—Russian *gypsum solonchak*, dominated by CaSO$_4$; *calcic solonchak*; and *nitrate solonchak* and *takyr*—as anthropogenic variants on long abandoned settlements. *Takyr* is a polygonal-patterned crust, highly saline and impermeable, often covering the sites of ruined cities in central Asia (eg Merva, destroyed in the thirteenth century and covering 225 ha). *Meadow and swamp solonchak* occur with high water tables on clays. *Calcic solonchak* occur in S.E. Alberta on

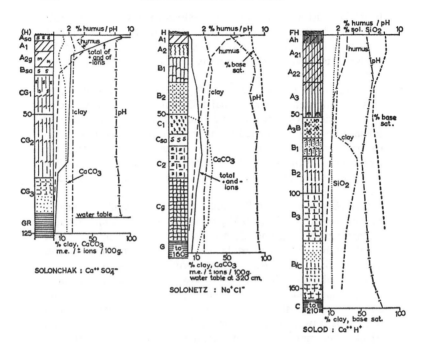

Fig. 8 Profiles of halomorphic soils from Rumania. Chemical and physical analyses are shown to the right of the profiles, many sharing the same scale. Note the curves for pH, lime content and clay. See text for other dominant ions. Location: Lower Danube (28°E); (after Florea)

calcareous marls, and in N. Dakota on saline glacio-lacustrine sediments, with Ah, Aca, Cca, Csa, G, Dg profiles.

Solonets have shallow neutral A horizons due to the leaching of soluble Na, K, and carbonates, from solonchak, caused by: (1) a falling water table on rejuvenation; (2) increased rainfall on climatic change, or (3) irrigation. The dark grey or brownish B1 horizons are then colloidally enriched and have a distinctive columnar structure, with rounded 'cauliflower' tops (see 7A, Fig. 13). Underneath is a lighter grey B2, deficient in carbonates. Only in the very light grey C1 horizon does lime suddenly increase (Fig. 8).[39] Thus apart from the saline Csa, most of the C horizon is a C2ca to c 1 m depth, followed at depth by a massive Cg horizon (Fig. 8, 2).

Soils where an acid front has penetrated to the base of the B at 50 cm may be termed *non-calcic solonets* and occur in moist climates. Decalcification is not so deep-reaching in drier areas, or on calcareous parent materials, and *calcic solonets* are formed. In wet periods the water table is from 0 to 1 m deep and the A horizons become pasty and peptized. On drying they harden or become powdery, and are subject to wind erosion. The columnar B1 is almost rock-like when dried, the water table then being at 2–3 m depth.

Black carbonate solonets form millions of acres of unproductive land in the Punjab, Sind, and the U.P. They are very deep (c 6 m) with a saline layer at 1 m depth. In W. Hyderabad, with 700 mm ø rain, leached Na, K, Ca and Mg sulphates and carbonates accumulate in depressions to form cryptosolonetsic '*phodas*' or quagmires.[41]

Solonets also result from seawater inundation—*maritime solonets*. They are common in Dutch and many other polders such as the Puglian coast of Italy or the Aegean of Greece, with pH 9·5, 1% salt content in the B horizon, and high (50%) Na saturation. On drying, the roots of plants colonizing the polders[42] release CO_2, which replaces the adsorbed Na by ion exchange:

$$2Na + CaCO_3 + H_2O \rightarrow Ca + 2Na^+ + 2HCO_3^-$$

Lime may also be removed from near the surface by hydrolysis and become available for ion exchange lower down; Na is then released and leached into the drainage waters. Yet this is not always possible for some irrigating waters are rich in Na and poor in Ca so little exchange takes place. The waters of the Colorado are fortunately poor in Na but rich in Ca, so that the danger of salinization is not so great.[37,43]

A firmer case can be made for the zonal status of the solonets than for the solonchak, for carbonate salinization or alkalinization is common in the aridisol-pedocal transition zone at 40°N and also in the desert-subtropical ferrisol transition at 30°–15°N. These are zones in which climatic change has recently led either toward enhanced, or to diminished, alkalinization.

Solods form the next step in desalinization and are termed 'salt-earth podzols'. They have a mull-like F layer and deep, weakly acid, grey-brown, sandy A horizons, which form at the expense of the 'domed' B1 of the solonets, by enhanced leaching. Next comes a deep B(G) horizon of compact clay, oxidized in its upper part as a B1–2 and akin to a planosol (Fig. 5, 8–9 and Fig. 8, 3).

Solods are more productive than solonets or solodized solonets; solodized solonets have peptized, potentially dusty and erodible A horizons[44] and excess Mg—a transitional form, known as *Mg solonets*.

Solods were first noted in 1926[45] as resulting from one of three causes: (1) leaching of solonets as a slow natural process; (2) repeated leaching of saline soils by irrigation with improved drainage or falling natural base levels; (3) anaerobic bacterial activity mobilizing Fe compounds in waterlogged soils or in those flooded by snowmelt, producing a *surface water solod*.

The most common solods are *steppe solods* with deep ground-water. *Meadow solods* occur if water is near to the surface. Gleyed, peaty and sod solods are reported from the Irtysh (Omsk) lowlands.

Solodized arid brown soils or *mallisols*[46] are widespread in South Australia and are compared to the red/brown aridic soils of the USA. Mallisols have marked Bca horizons, increasing fineness of texture and alkalinity with depth, and high contents of soluble salts in the C horizon. A profile sequence would be A; AB (pH 7·5–8·8); a prismatic B1–2 (pH 8–9); light brown hard Bca (pH 9); very light grey soft Cca (pH 9·5), and a Csa with pH >9·5.

Many other saline and alkaline soils, salt 'pans' and crusts are known in aridisols, and irrigation with saline water in arid areas soon produces saline horizons, as in the irrigated basins of Iraq and W. Pakistan.

1 E. Mückenhausen, 1960, profile 53

2 For a discussion of hydromorphic soils see Chapter 7 of S. A. Wilde, *Forest Soils*, 1958

3 E. Crompton, JSS, 3, 2, 1952, 277–89

4 C. A. Bower and M. Fireman, *Saline and Alkaline Soils*, in USDA Yrbk., 1957, 282–90; 4b D. H. Yaalon, Bull. Res. Cncl. Israel, 11G, 1963, 105–31

5 G. Brown, JSS, 4, 1953, 220–8 and 5, 1954, 145–55

6 C. Bloomfield, JSS, 2, 1951, 196–211

7 J. E. Dawson, *Organic Soils*, Adv. Agron. 8, 1956, 377–401

8 H. Sjörs, Endeavour, XX, 80, Oct. 1961, 217–24

9 N. W. Radforth, in *Soils of Canada*, Ryl. Soc. Can., Spec. Pub. 3, 1961, 115–39

10 B. van Heuvelen, TICSS, 7, 1960, V. 27, 195–204

11 E. J. Russell, *World of the Soil*, p. 259 and Plate XXI

12 W. W. Weir, Hilgardia, 20, 1950, 37–56

13 J. R. Gasser, JSS, 12, 2, 1961, 234–41

14 I. I. Karmanov, Sov. SS, 8, 1960, 828–33

15 J. W. O. Jeffery, JSS 11, 140–8 and 12, 172–9, 1960–61

16a H. Fastabend et al., Geol Jb., 78, 1961, 139–72 and
16b H. Fastabend, Geol. Jb. 79, 1962, 863–6

17 C. H. Edelman, *Soils of the Netherlands*, 1950

18 E. Mückenhausen, 1960, profile 45 (Grey Plaggenesch) and 46 (Brown Plaggenesch)

19 H. Zakosek, ZPDB, B, 93, 1, 1961, 38–43

20 E. G. Hallsworth et al., JSS, 6, 1, 1955, 1–31

21 C. H. Edelman and R. Brinkman, SS, 94, 1962, 366–70

22 F. Scheffer et al., ZPDB, 98, 1, 1962, 1–17 and 90, 3, 1960, 18–36

23 J. J. Reynders, NJAS, 9, 1, 1961, 36–40

24 I. A. Zvorykin and N. Katakousinos, *A Study of the Soils of Thessaly*, Inst. Chem. and Agric., Piraeus, 1960, 1–118 (p. 112)

25 W. L. Kubiena, *Soils of Europe*, 178–91

26 B. W. Avery et al., JSS, 10, 2, 1959, 177–95

27 E. Schlichting, ZPDB, 58, 1952, 97–106

28 E. Mückenhausen, 1960, profile 8

29 J. Fink, Mitt. Geog. Ges. Wien, 100, III, 1958, 92–134

30 J. M. Albareda Herrera et al., Ann. Edafol. Agrobiol., 20, 1961, 233–63

31 See *Soil*, USDA, Yrbk, 1957, refs

32 J. C. Frye et al., *Gumbotil, Accretion Gley and the Weathering Profile*, Illinois St. Geol. Surv. Circular 295, 1960, 1–39

33 J. C. Frye et al., AJS, 258, 1960, 185–90

34 R. B. Grossman et al., PSSSA, 23, 1, 1959, 65–75

35 N. J. Yassoglou et al., PSSSA, 24, 5, 1960, 396–407

36 R. Maréchal and R. Tavernier, Pédologie, VII, 1957, 199–203. See also M. G. Cline et al., PSSSA, 27, 3, 1963, 339–44

37 L. A. Richards, *Diagnosis and Improvement of Saline and Alkaline Soils*, USDA, Handbook, 60, 1954, 160 pp

38 A. D. Ayers et al., SS, 90, 2, 1960, 133–8

39 N. Florea and V. M. Fridland, Solurile, in *Monografia Geografica a Republicii Pop. Romine, I, Geografia Fizica*, 1960, 463–540. On nitrate solonchak see: V. A. Molodtsov, Sov. SS, 6, 1961, 659–63

40 V. A. Kovda, *Origin of Salinized Soils*, 1–2, Izd. Ak. Nauk. SSSR, 1946–7. On Hungarian 'szik' see: A. Arany, Al. Thaer Arkiv, 4, 1, 1960 23–36

41 P. G. Krishna et al., SS, 70, 1950, 335–44; E. N. Ivanova, Poch., 4, 1963, 20–9

42 Studies of the leaching of saline soils are given in *Reclamation of Salt-affected Soils in Iraq*. Intl. Inst. Land Reclam., Pub. 11, 1963

43 J. O. Goertzen and C. A. Bower, PSSSA, 22, 1, 1958, 36–7

44 W. K. Janzen, JSS, 12, 1961, 101–10 and 13, 1962, 116–23

45 K. K. Gedroiz, *Alkali Soils: Origin, Properties and Improvement*, Bull. Ag. Expt. Stn. Nosovka, 44, 1926

46 K. H. Northcote, TICSS, 6, 1956, V. 2, 9–19

12

Desert and tundra soils

MANY desert areas show a sharp contrast between saline and aridic soils; the alluvial plains of Iraq[1] and Egypt[2,3] are examples. The Egyptian plains remain fertile with alkaline patches, while the Iraqi plains are excessively saline, for the Nile waters have little carbonate and derive inactive silt from tropical areas. The Euphrates, rising in N.E. Turkey (40°N), provides a pale brown alkaline alluvium; the Tigris, rising in S.E. Turkey (38°N) and in Kurdistan and Zagros (35°N), provides a grey-brown alluvium, rich in Ca- and Mg-carbonates.

South of Baghdad (32°N), between the two rivers, are three landscapes: (1) the subrecent levées; (2) the basin landscapes near both rivers; and (3) the intervening desert. On silty levées adjacent to irrigation channels the soils are non-saline and cultivable, and deepen with age. The main river levées are also silty, and homogenized by date palm and garden cultivation. The anastomosing lower-lying, older natural and artificial channels show saline soils developed in fine silt overlying basin clays. These soils (2) have been degraded by salt from irrigation waters and by irrigation leaching the levées; 'a more striking contrast than that between the fertile date palm orchards with their undersown crops and the barren saline basins nearby is hardly conceivable'.[1] The deserts of the central area (3) have dunes and sheets of pseudosand—clay flocculated by salt. Underneath this recent layer are desert soils with compact B horizons—*dried solonets*—the drifted sand being the erosion product of the former A horizons. All the soils of this area in Iraq form associations with a micro-landform basis identical to the site-geomorphology of Dutch alluvial plain soils.[1]

The soils of the Nile delta (31°N) and valley vary from *Nile clay*—*solods* and *solonetsic soils*—to soils greatly changed by man over long periods of time and termed *anciently irrigated alluvial silty meadow soils*.[3] They have increased organic matter and a developed surface structure. Older eluviated Nile terraces have more gravelly or sandy parent materials.

In the Qarun Lake-Faiyum Oasis area (29°N) dry saline and gypseous clayey soils occur on alluvial clay, and shallow calcic desert soils with sandy solonchak on the oasis borders. Irrigation with artesian water in the Kharga and Dhakla oases (25°N) has produced artificial organo-saline soils on lower slopes and dense uncultivated solonchaks in the basin. On the higher cultivated slopes on the oasis margin, red clay-loams occur on sandy calcareous alluvium, shallow red-brown loams on limestone, and pale yellow or brown soils on sandstones.[3]

Soil landscapes of alluvial and oasis areas contrast greatly with those of adjacent desert areas. Desert soils are mainly sandy; the coarsest sand is in the upper layers and finer sand at depth, with occasional silt and clay.[4] Other features are surface crusts and gravel layers (reg).[5] Local differences in aridisol profiles are textural; the main zonal differences are in colour and dominant salt type. Many desert soils in Australia, the Sahel and southern North America are red; the cooler Russian and Argentine deserts have light grey *serozem*. Relic braunlehm are also known.[6]

Serozem are poorly structured, alkaline-calcareous AC soils, rich in silt (c 70%). They have little or no humus and a scattered xerophytic plant cover. A profile sequence would be A1, A2, A/Ca, Ca, C.[7,8] The A2 may be darker than the A1, with <2% humus; the Ca either calcipan, or spongy lime above a loose C horizon. Serozem may be associated with *grey solonchak* in depressions, and with *paraserozems* on coarse sands. In Turkestan (44°N) serozem are related to a series of terraces or piedmont steps of differing age. On the lowest, youngest, surfaces *light serozem* occur, with a rainfall of 120 mm and 14°CØ. On higher, older surfaces *typical* and *dark serozem* occur, with 320 mm rain (13°CØ) and 460 mm rain (12°CØ) respectively. In Egypt (30°N), serozem are light cinnamon-brown and carbonate-rich—a *subtropical serozem*.[3] The same colour is noted in the semi-arid grass savanna of S.W. Africa (22°S) with

150–400 mm rain. Here the true desertic sands are red with skeletal calcareous crusts.[8]

The red subsoils of Australian *stony desert tableland soils* reflect the remnants of former, now truncated, soils developed under pluvial tropical conditions, and they have stony brown A horizons.[9] Other Australian aridisols[9] are the *desert loams* of S. Australia, powdery *red calc loams or clays* on the limestone of the Nullabor Plain (31°S), and the *red hardpan soils* with stony *gibber* surfaces and vesicular structure in W. Australia. In Northern Territory, *reddish desert sand-plain soils* are most common.

Soil landscapes also vary greatly within deserts. Apart from bare rock, pebbly *reg*, dunes and sand fields—*erg*—there are relic soils and soils related to climatic sub-zones of varying, yet low, rainfall. For example, N.E. Africa has three zones[3]: (1) along the Mediterranean coast (32°N), rainfall of 30–75 mm Ø gives *plateau serozem* with incipient salinization; (2) south of 30°N, soil formation is lacking, except in oases. *Gypseous crusts* form and salts accumulate in wadis. (3) With no rain and low humidities *abiotic deserts* occur, as in S. Egypt (20°–27°N). The upper rock layers are cracked and loosened, change colour and form structural units under thermal influence; salts and gypsum accumulate on the surface of weathering igneous rocks and in fissures as crusts, while siliceous rocks show silica crusts.[3]

The soils of American deserts also vary. Valleys in the Colorado Desert have alluvial and saline soils of varying texture; mesas have sandy mixed soils; pediments have cemented gravelly or coarse sandy soils of a colluvial nature, with pH c 8·5, and stony pavements are common, the combined result of upward movement of stones and of deflation.[10]

Soils and surface features in tundra areas

The tundra and arctic is the other major zone in which soil development is minimal. Here, though heat is lacking, there is no deficiency of moisture despite low precipitation, and chemical weathering is weak.

Site is the main influence in tundra soil development, with impeded drainage in lowlands and incipient leaching on elevations. Soil formation is also influenced by the thickness of the active layer, while the underlying permafrost—a state of ground, not a parent material—is a DF horizon (Fig. 14, 8).

The depth of the summer thaw varies from 30 cm at the Arctic Circle in a maritime climate (68°N, 60°E), to 1 m on peat bogs, and 2 m on sands at Lake Baikal (53°N, 106°E). Saturated anærobic conditions obtain in summer, with the thawed layer a bluish-grey colour and in a viscous, plastic or thixotropic state. Organic and mineral portions may be mixed together and movement take place.[1] An organic layer may form at the base of the active layer, resting on the permafrost; the origin of this Hb layer is in dispute.[2]

As the active layer retards refreezing, it becomes trapped between the permafrost and the thin refreezing surface. It is subjected to lateral pressures and contorted to form *cryoturbate* features.[3] Such folded (turbate), contracted, fractured or flow features are found, fossilized, in old periglacial deposits in central Europe.[4]

Flow of thawed material occurs on slopes >2°, forming accumulations of head, combe-rock, taele gravel and fans—all parent material for later soil formation, though often overlain by thin layers of colluvium. Flow also exposes new permafrost for thawing in the area of negative balance of denudation.

Polygonal soils of stone rings surrounding fine material form on drier elevated areas with slopes <4°—often, too, outside the tundra zone on elevated cool oceanic sites.[5] Such polygons become elongated at 3°–7°, forming stone stripes at 7°–26°. Sørensen[6] classified high-arctic soliflual soils into those of inhomogeneous material (stones and blocks with some fines) and those of stoneless fines. These textural groups form different soils according to four variable climo-sites and slope angles:

TABLE 11

CLIMO-SITE AND TEXTURE IN ARCTIC SOIL PATTERNS[6]

Climo-site	*Texture*	
A Little snow cover	Stony + fines	Homogeneous fines
1 Dry, poorly watered	Talus and small stone nets	Cell soils
2 Well watered, summer dry	Stone stripes or polygons	Stone-free polygons and loamy hillocks
3 Wet sand	Block terraces, stone stripes or garlands on slopes. Polygons on flats	'Crater soils' of pitted flats
B Deep, long-lasting snow and intensive thaw		
4 Wet sites, some standing water	Plaster swells or *slud*	Loamy swells and mudflats

In addition, steep slopes on homogeneous fines would show: (A1) earth tongues; (2) earth glaciers; (3) earth streams and (B4) mud flows.

Tedrow considers that mechanical weathering is overemphasized in arctic soils.[7] Most developed arctic soils form on well drained regosolic material where plants provide acids and hydrolyze minerals. Typical sites are ridges, eskers, river terrace margins and dunes on which *arctic brown soils* develop. Such soils have shallow H, Ah, A2, A3C, C profiles with 10% organic matter in the Ah; a brown, acid (pH 4), crumb-structured, root-rich, A2; a single-grained, yellow, A3C: the C is olive-brown, neutral to alkaline, fine textured, stony, and with some carbonate deposition.[8] Often fine loessic material is added to the surface (Fig. 5, 24). Fe and Al are partially mobilized from the surface, indicating a weak podzolic process.

With increasing hydromorphism, arctic brown soils progressively form *eluviated tundra, meadow tundra, tundra gley, halfbog*, and finally *bog soils*. The sequence: *lithosol, arctic brown, nano-podzol, typical podzol* is possible; Russian workers also regard the arctic brown soil as divisible into typical, calcic and podzolic forms.

Podzolization in tundra soils is regarded as an intrazonal form on coarse materials, noted as such as far north as 68° in the Kola Peninsula, and at 67°N in S. Strømfjord, Greenland,[9] on weathered Archaean gneiss in the outer-fjord (maritime) area. At the fjord head on north-facing slopes, acid (pH 5) *sod brown arctic soils* occur, with basic soils (pH 6·4–8·5) on south-facing slopes. In sheltered hollows *alkaline arctic soils* occur, with pH 8·5–9·2. Summer capillary currents and 100 mm Ø precipitation are not the primary cause, for the soils are sited on dried-up lagoons formed by recent isostatic uplift.

The most widespread arctic soil is the *tundra gley*,[7] developed on gently undulating valley slopes, flat watersheds and valley floors. It is a surface water gley with a peaty surface and medium textured mixed 'parent materials'—the thawed layer.

Tundra gleys are <0·6 m deep, the Ah1–2 and A2 layers are y/b to brown if slow lateral drainage is possible to lower sites; but dark grey in low-lying *meadow tundra*. Lower horizons are olive-grey, with ochreish streaks, and are acid. Hydromorphic variants have *semi-bog* profiles with 20 cm-1 m peaty surface layers; full *bog peats* having 1-6 m deep organic layers.

In S.W. Alaska (62°N)[10] and in W. Siberia the frost table is at some depth and does not affect soil formation in summer. Here deep *meadow tundra* and *sod glei* soils develop. In rocky areas, in Labrador, and on the Brooks Range and in the E. Siberian Mountains, skeletal soils form on dry permafrost.

Studies on Bolshevik Island (78°N, 103°E) typify the pedologic environment of the extreme arctic,[11] with ungleyed *low humus arctic sod soils* on the maritime plain, with much free iron in the A, and an alkaline soil solution; *humified sod arctic soils* on south-facing warm 'continental' slopes, with a more acid A horizon; and *rubbly mountain arctic soils* on moraines near to glaciers. Positive temperatures obtain for only two months and frosts occur throughout the year. A summer precipitation of 60 cm is not supplemented by snow thaw, for this runs off over frozen ground. There is high evapotranspiration in summer, with high winds and continuous insolation, which inhibit deep leaching.

Though many variants of tundra soils exist, there is little point in considering them as zonal. Soils rarely show developed horizons, being mixed by cryogenic processes. Most are gleyed or regosolic; some are humic, most are acid, and a few are saline. Arctic brown soils and podzols are local exceptions. Processes of soil formation seem only to operate in the short summer season and disruptive cryogenic processes dominate.

Desert soils
   1 P. Buringh and C. H. Edelman, NJAS, 3, 1, 1955, 40–9
   2 W. Schoonover et al., Hilgardia, 26, 1950, 565–96 and M. M. Elgabaly et al., JSS, 13, 2, 1962, 333–42
   3 A. N. Rozanov et al., Soils of U.A.R., Sov. SS, 5, 1961, 572–5
   4 M. Rim, JSS, 2, 1951, 188–95
   5 C. C. Nikiforoff, SS, 43, 1937, 105–31
   6 W. L. Kubiena, Erdkunde, IX, 1955, 115–32
   7 R. Ganssen, Die Erde, 1, 1960, 15–31 and ZPDB, 94, 1961, 9–25. Also E. V. Lobova, Bull. Ass. Fr. Ét. du Sol, 5, 1960, 269–82
   8 A. N. Rozanov, *Serozems of Central Asia*, IPST, 1961 and Poch., 1952, 611–28
   9 C. G. Stephens, *A Manual of Australian Soils*, 2nd ed., 1956, CSIRO, and H. C. T. Stace, *Morphological and Chemical Characteristics of . . . the Great Soil Groups of Australia*. CSIRO, Adelaide, 1961, 209–30

10 M. E. Springer, PSSSA, 22, 1, 1958, 63–6

Tundra soils

1 Information on permafrost and the thawed layer is profuse—reviews include: R. F. Black, BGSA, 65, 1954, 839–56; S. Taber, BGSA, 54, 1943, 1433–1558; P. J. Williams, GJ, 123, 1957, 42–58; E. A. Fitzpatrick, JSS, 7, 1956, 248–54 and SGM, 74, 1, 1958, 28–36; A. Jahn, Uniw. Wrocklaw. Nauk. Przyrod. B, 5, 1961, 1–34

2 J. R. Mackay, Dept. Mines, Ottawa, Geog. Paper 18, 1958, 1–21. See also N. A. Karavaeva and V. O. Targullan, Poch., 12, 1960, 36–45

3 On cryopedology see: K. Bryan, AJS, 244, 1946, 622—42 and A. Cailleux and G. Taylor, *Cryopédologie*, Expdt. Pol. Fr., Paris, 1954

4 H. Liedtke, Wiss. Zeit. Humb. Univ. Berlin, Naturhist. Rke. VII, 3, 1957, 359–76

5 A. L. Washburn, BGSA, 67, 1956, 823–66

6 T. Sørensen, Medd. Grønland, 93, 4, 1935, 1–69. This has a good bibliography of early work on arctic soils

7 A series of papers: with J. V. Drew et al., TAGU, 39, 1958, 697–701; with J. E. Cantlon, Arctic, 11, 1958, 166–79

8 With H. Harries, Oikos, 11, 2, 1960, 237–49; SS, 88, 6, 1959, 305–12 and SS, 80, 2, 1955, 265–75

9 T. W. Bøcher, Medd. Grøn., 147, 2, 1949, 33–64

10 S. A. Wilde and H. H. Krause, JSS, 11, 1960, 266–79 and J. C. F. Tedrow et al., JSS, 9, 1958, 33–45; also SS, 80, 2, 1955, 265–75

11 I. S. Mikhalov, 'Bolshevik Island', Sov. SS, 6, 1960, 649–52 and N. M. Svatkov, 'Wrangel Island', Poch., 11, 1958, 52–9

13

Light coloured soils of the boreal zone (podzols)

THE zone of the boreal forests is considered coincident with the zone of podzols. This is, however, provided that parent materials, site and man permit, for other soils can form of a non-podzolized nature. Large areas of the alleged podzol zone have regosols, gleys, brown earths and peats. Peats are widespread; a climax subzonal form south of Hudson Bay, in the west Siberian Lowland, central Alaska, Labrador, the Bothnian lowlands and in Kamchatka.[1] Thus *boreal peat and sod soils* form, especially in mid-continental permafrosted parts of the taiga, while *peaty podzols, peaty gley podzols* and *podzolic gleys* are common in oceanic areas. Gleys and solods are also found in mid-continental areas; thus podzols are common only on well-drained sites.

Russian soil scientists divide this *zone of podzols* stretching across their enormous country both latitudinally and longitudinally into a series of sub-rectangular compartments.[2] The four latitudinal belts are the subarctic-, northern-, central- and southern-boreal sub-zones, each with distinctive soil processes. These subzones are then divided longitudinally into five sections, the western oceanic, west-, mid-, and east-continental, and the eastern maritime (cool monsoonal) sections. Mountains and permafrost vary the scheme and the compartments vary in size. Further sources of variation in each are parent materials, vegetation and length of human occupance.

Thus 'boreal' and 'podzol' are by no means synonymous and podzolization is, in fact, highly variable in its effects. It is man-induced in oceanic parts of north-west Europe,[3,4] a zonal process only in high mountains and on well-drained parts of the taiga,[3,5] and intrazonal on coarse materials in tundra and tropical latitudes.[6]

Many other soil forms of slightly 'podzolic' appearance exist in warm temperate areas as well as within the boreal zone. There are, indeed, five stages if a podzolized soil is to form: (1) *decalcification* of parent materials; (2) formation and migration of clay complexes (*illimerization*) or *lessivage* (Fig. 9, 2); (3) formation of free oxides by intense weathering of silicates and clay destruction (*silicification*) (Fig. 9, 3); (4) translocation or *cheluviation* of Fe and Al by acid organic matter (Fig. 10, 3); and (5) *reprecipitation* as a Bs layer.[7]

Podzolization therefore implies an intense, though usually shallow, alteration of the upper mineral soil by organic acids, and various trees—pine, fir, larch, willow, spruce and birch—produce litters of varying efficiency in cheluviation.[8]

Dokuchayev's original purpose in using the term *podzol* was to limit it to a shallow 'ashen' surface layer,[9] which we now know is best achieved in cool temperate humid climates on psammitic materials or acid igneous rocks, by cheluviation under mor and where microbial activity—apart from that of fungi—is low. Dokuchayev's notion has, with the course of time, been broadened to describe a whole sequence of horizons: LFH, Ah, A2, Bh, Bs, B22-3, C1-2. Distinct LFH horizons—the *forest floor* or *debris layer* —vary in combined depth from 1 cm to 1 m of highly acid, N-deficient mor. It is itself a zonal irreversible form in north Sweden and western Siberia,[10] and it is azonal in oceanic and in warmer continental areas (eg S. Urals); the result of planting conifers and it is reversible by planting birch.[11] The mor is sharply separated from the mineral A horizons, which consist of grey, pale brown or white (bleached) sand that is highly acid and from which Al and Fe oxides have been removed. In some, the upper A horizon has an admixture of fine humus, or debris of decayed roots—an Ah of similar colour to the mor. Thus the A2 is most distinct, with a single grained A21, and an A22 with an ashen or cinder-like pseudo-structure, for colourless organic colloids cause adhesion of quartz in wet soils, to form the ash-like '*podzol*'[9] (Fig. 5, X, Y, Z).

Underneath the shallow (<0.5 m) eluviated horizons (Ae) is an upper illuvial horizon, which may be deep reaching and soft (c 0.5 m —*orterde*), or else shallow and hard (<5 cm—*ortstein*). These horizons are known as Bh if dark brown and illuviated with C-rich humus, and as Bs if reddish-brown and enriched with Al and Fe oxides and colloidal Si.[12] If the Bh is present alone on iron-deficient material a *humus podzol* is formed; a Bs layer alone gives an *iron podzol*, while, if both are present, an *iron-humus podzol* is formed, in which case the Bs is richer in colloidal Si and may have 20% base saturation.[13] (See Fig. 9, 5.)

The total depth of the A and B horizons varies more with parent material and site than with climate and vegetation. LFH are thin and Ah rare on well drained sites; on wet sites the Ah is better developed, accompanied by thicker LFH.[14] Ortstein is also thin or absent on dry sites on coarse material, being more common on finer-textured or ill-drained soils with periodically high water tables (Fig. 5, 23). On slowly drained shallow soils ortstein forms upon or within slightly weathered rock.[15a] Both A and B horizons are thicker on sands than on loams or slightly glauconiferous or calcareous sandstones.[16] Variations in the depth of podzolized layers may be shown by means of isopachytal maps.[15b]

Podzol profiles are best explained by variations around a modal type, with thick, thin and minimal profiles.[17] In the *minimal podzol* the A horizon is <1 cm deep under newly planted pine, beech or other acid litter, a *micro-* or *proto-podzol*; and the B horizon is the lower A or upper B of some other pre-existing soil, there being little or no Bs. The *thin podzol* is either a zonal form or a local site type, with 5 cm bleicherde on a deep B horizon, and orterde tendencies. The *modal podzol* has *c* 10–30 cm bleicherde, about two-thirds the depth of the 20–50 cm deep Bh-s orterde (Fig. 10, 2). Finally the *deep strong podzol* has at least 50 cm bleicherde, of equal or greater depth than the B horizons (Fig. 9, 6). An *ultra-podzol* is also known,[15a] with 1 m deep A, and vestiges of Bs, most of the sesquioxides having been laterally eluviated to the drainage waters.

Hydromorphic variants of the podzol are caused by waterlogging of the A layers on flat sites if the Bh-s are impermeable (pan); though more frequently the high water content of a thick hydrophilic orterde (or the Bt of a *prepodzol*) (Fig. 9, 4) will give moisture impedence and produce pseudogleying, forming a *pseudogley-podzol*.[18] Compact *molken podzol* form on fine grained silt- and sandstones, as well as on older erosion surfaces and peneplains in

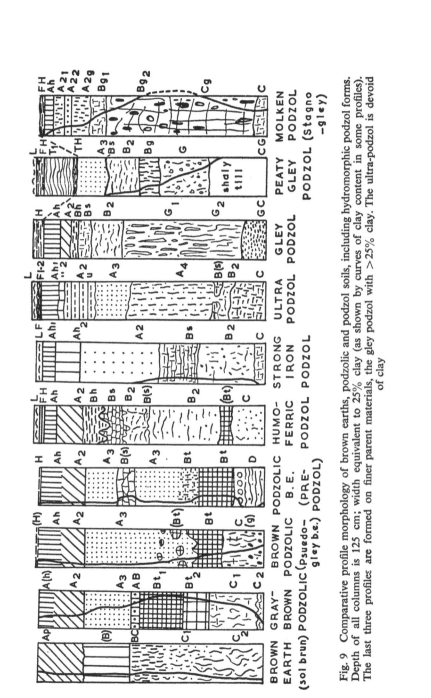

Fig. 9 Comparative profile morphology of brown earths, podzolic and podzol soils, including hydromorphic podzol forms. Depth of all columns is 125 cm; width equivalent to 25% clay (as shown by curves of clay content in some profiles). The last three profiles are formed on finer parent materials, the gley podzol with >25% clay. The ultra-podzol is devoid of clay

Hercynian uplands (Fig. 9, 10).[19] More extreme hydromorphic forms
are *gley podzols* (with waterlogged B3g, G1–2) and *peaty podzols*
(Fig. 9).[20]

Oceanic and boreal podzols may now be compared. Boreal
podzols are a zonal bioclimatic effect; oceanic podzols follow on
lessivage, forming within a deep sesquioxide-rich A2 which is then
cheluviated by conifer debris, or losing bases if deciduous trees are
felled and heath replaces mull with mor.[34] Lyford[17] has presented
the difference another way, showing that there are two kinds of
modal podzol subsoil: (1) no greatly different B2 beneath the
orterde—a red coloured layer, as in the grey-brown podzolic and
r/y podzolic soils; (2) apparently polygenetic soils, with a bleicherde
over a brown B2 horizon, due either to climatic change or vegetation
clearance, which would give complete decalcification and a stop to
the supply of base-rich litter.

For the oceanic (Atlantic) environment Kubiena[21] has many subdivisions of podzols. Northern (subarctic) *cryptopodsols* and *molken
podsols* are best developed on Bunter and Jurassic sandstones or
upon fossil *plastosols* (red and brown loams), in central Europe.
The *humus podsol* is most common on heathlands surrounding the
North and Baltic Seas, usually on Saale till or Weichselian fluvioglacial sands (Fig. 5, 15). The *iron humus podsol*, with thin ortstein,
is typical of the northern heaths of Sweden and Finland (Fig. 10, 3)
and of dwarf shrub areas above the tree line.[22]

Modal *iron podsols*[23] are the true forest podzol, usually under
spruce, with deep F1, 2, 3 layers, and deep orterde rather than pan.
They occur widely on acid igneous rocks in central Europe, in
freely drained sites in eastern Scotland, in the western Alps, and on
eskers in southern Finland. There may be some humus patches or
streaks in the deep B(s). All the forms and areas of podzols outlined
are not mutually exclusive.

Other more limited podzol forms are *alpine sod podsol*, the *dwarf-*
or *nano-podsol* of the sub-arctic and of dry mountains (Fig. 15),
and *vestigial podsols* on gravels, with slight iron films at great depth.

Podzol development in boreal Siberia[5] and in mid-continental
Canada[24] contrasts greatly with that in oceanic areas. There is
lower rainfall, with a summer maximum, long winters and permafrost—all preventing deep leaching.

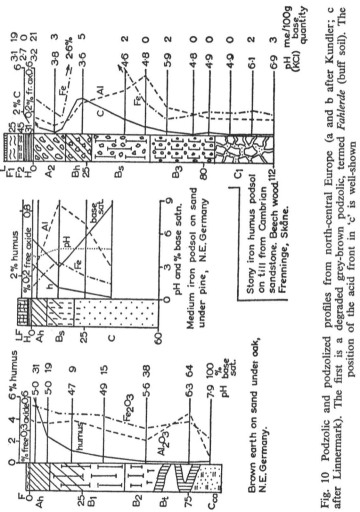

Fig. 10 Podzolic and podzolized profiles from north-central Europe (a and b after Kundler; c after Linnermark). The first is a degraded grey-brown podzolic, termed *Fahlerde* (buff soil). The position of the acid front in 'c' is well-shown

Reference to the world map of soils[2],[25] shows that at 30°E the zone of podzols stretches latitudinally from 68°N (Murmansk) to Kiev (51°N). From Lake Inari to the Usa valley (66°N, 58°E) the permafrost limit almost coincides with the northern limit of podzols. At the longitude of Moscow (38°E) the zone narrows from 67° to 55°N, and extends east to the Yenisei River (90°E), though the plain of West Siberia is edaphically wetter than the European area west of the Urals.

Further east, in eastern Siberia[26], *mountain forest podzols* dominate, from 72° to 50°N, with dry permafrost in the north and chernozemic soils in the south, near to the national border.[27] Yet E. Siberia has a sixfold relief division: (1) the Central Siberian Plateau (<1700 m high); (2) the Irkutsk-Baykal Mountains in the south-centre (<3000 m); (3) the Lena basin; (4) the N.E. Siberian mountains (to 3200 m); (5) the eastern coastal mountains of Sikhote Alin and Kamchatka; (6) the Amur basin.

Latitudinal sub-zonation of each of these sections[2] is achieved by reference to (i) increasing tendency to tundra and active-layer features in the north; (ii) solodization in the south-east and upon recent parent materials; (iii) less plant debris in the south, and less acid, being derived from deciduous trees; and (iv) more intensive cultivation in the south, mixing the A horizons and producing a 'sod' or turfy upper layer. In warmer areas, too, the nature of the parent material has more influence in soil development.

Within the *western part of the Russian zone of podzols* a line may be drawn parallel to its southern boundary from Kaunas to Berezniki (59°N) to delimit: (a) the *gley-podzolic* and *peaty podzolic* soils to the north and east of Leningrad; (b) the *derno-(sod) podzolic* soils to the south and west and (c) the drier *iron podzols* in the centre and east. There are exceptions to this scheme: the Pripyat marsh-lands and the bogs north of Smolensk; the limestone terrain of Estonia and Vologda with *rendzinic* soils; and the recent Würm drifts of western Latvia and Lithuania with *brown earths*. Areas (a)–(c) are taken in turn:

(a) Western Komi, Archangel, and the lowlands of Finnokarelia form a sub-zone of *gleyed podzols*, though without permafrost.[28] On wet sites, *peaty podzolic gleys*, with T, A2, Bhg, Cg profiles develop into *podzolic bog* soils, which are highly acid, have 1 m deep T, of sphagnum-hypnum peat, and are thixotropic.[29]

(b) *Sod podzolic soils* are rather a vague entity, possibly equivalent to the gray-brown podzolic in America and the sol brun lessivé in Europe, or an intergrade to the brown podzolic soils; for the Bt show signs of clay destruction and there are other differences, including a yellowish (*palevo*) A2 or A2B.[30,31] They form under mixed forest, their latitudinal limits at 25°E are 53°–59°N, at Moscow, 55°–57°N, trending north of the Volga from Gorkiy to Kazan, and then north of the Kama to Perm (56°–58°N). Some consider their podzolic appearance without Bhs layers to be a 'zonal' climatic form;[32] others a result of long cultivation or rapid turnover of neutral litter giving a firm mull.[33] The climate is such that the topsoil is frozen in winter and wettest in spring and autumn; soluble Fe^2 is therefore oxidized and rendered insoluble in summer and is not easily leached. Microbial activity and some earthworms exist which, with roots of crops and grass, form stable humus[30]. The Russians term the whole a *sod-cultivation process*, which causes an even distribution of nutrients and humus in the A horizons and inhibits cheluviation. These soils are termed *glossudalf* in 7A.

(c) *Iron or forest podzols* have shallow Ah and A2 horizons of a light grey colour; the B are yellowish orterde. Thick F layers impede aeration, the acid humus sols having a reducing effect. The type area would be Kostroma-Kirov, extending westwards on sandy parent materials into the sod-podzolic zone 'as far south as Smolensk'[21] and into S. Finland in Mikkeli Province on stony moraine. The Urals from 56° to 53°N have thin *mountain forest podzols* (nano- and neo-podzols) and *upland forest brown* and *grey forest soils* to 52°N.

East of the Urals, in *West Siberia*,[34] the southern boundary of permafrost moves sharply southward, traversing the plain at *c* 62°N. With an impermeable frozen substrate, the supply of water to plants is increased and grass growth in summer is dense and coarse, with root systems near to the surface. Organic horizons are therefore deep and rich, and humus tongues penetrate into frost fissures. Often mineral A horizons are very dark coloured and shallow, for cheluviation is attenuated by permafrosted subsoil. A *pale yellow sod-forest soil* forms in the central sub-zone (56°–62°N); with *surface-gleyed podzolics* or *humic gleys* in the northern taiga, before the wooded tundra is reached at 66°N. The wooded tundra has hummocky frost forms, *tundra gleys*, and extreme soil thixotropy in the waterlogged lands of the Ob Gulf.[2] *Peaty hydromorphic soils* are common between 56° and 62°N, while the valleys of the Irtysh, Ob and Yenisei have *meadow soils* in belts extending for 50 km to either side of the rivers.[27]

Next southward in latitudinal succession (*c* 55°–56°N) is a belt of *grey forest soils* and *meadow chernozems* trending due east from Chelyabinsk to Krasnoyarsk,[35] matched in the southern Urals by mountain steppe soils. Grey forest soils also occur in European Russia[36] to the north-west of the degraded chernozem and are long cultivated, resembling the brunizem of USA, while, in western Siberia, they show more clearly the long continued effect of invasion of chernozemic soils by birch and larch, and have only recently been widely cultivated.[27,11] In W. Russia the soils are *light greyish brown forest soils*, certainly more grey than in western Europe, with a rainfall of 500–700 mm Ø. They also occur in N. Bulgarian uplands with thick B horizons—Ap, A2, B1–4, C.[37]

Grey forest soils have thicker organic horizons than sod podzolics, lack a grey-brown A2, which, though weakly acid, has 80% base-saturation, and have H, Ah, A2, AB, Bt, C, Cca profiles—a nutty structured A2 and mycelial lime in the Cca. They may be sub-divided[38] into *dark grey forest soils* on loams, on level land, and mainly in the east, with thick organic layers, 10% humus, and neutral pH (6·5), with hydromica dominant. *Typical grey forest soils* have *c* 5% humus; and dominant clays are hydromica and kaolinite. They are best developed on uplands in the Siberian chernozemic zone. *Light greyish brown forest soils* have 3% humus and marked kaolinite dominance, podzolic tendencies are evident due to higher rainfall and less intense subsoil freezing. Such soils occur west of the Urals, being widespread in the Tatar ASSR along with podzolic soils, also in western Ukraine and in the Siret valley of E. Rumania. The potential productivity of grey forest soils is limited by their unstable surface structure, which is easily eroded by wind or raindrop impact, exposing the fine textured B at the surface, which horizon shows prismatic structure and 'clay skins'.

The central Siberian grey forest soils occur as islands among the degraded chernozem which are at lower levels.[27] For example, on the left bank of the Yenisei, north of Krasnoyarsk, is a wide undulating piedmont with forest steppe.[35] *Grey forest soils* occur on watersheds and hills, *leached chernozem* on lower terraces on younger loams and clays. Varying development of bleached A21, gleyed Blg, and thickness of surface humus depend on elevation or on water table depth, in turn determined by geomorphic layout.[35]

The origin of *grey forest soils* is explained by their subzone being warmer, drier and unforested in mid post-glacial times. A later, cooler, more humid climate caused invasion of forest on to the steppe, greatly degrading the already leached chernozem.[11,35] Amelioration within historic times has started to reverse this degradation, aided by cultivation, instigating a 'sod process' or *progradation*, with bases adsorbed in the A layers. The process is due to a general bioclimatic change and has been noted in many areas— Kansk (97°E), Tomsk (85°E), the S. Urals (58°E) and Kuznetsk (46°E). Invasions of *populus* on to chernozemic soils in Canada have caused degradation to *gray wooded soils*,[39] though these are not precisely the same as Russian *grey forest*. The later progradation is not documented.

In central Siberia, the Soviet land stretches far to the north (78°N) including Severnaya Zemlya. *Moist tundra and subtundra soils* here reach *less* further south than in west Siberia to be replaced by the *mountain tundra* of the Putoran Mountains (68°N) and the *nano-podzols* of the Gory Plateau.[2] Mountain tundra soils also occur at 800 m on the Yenisei ridge (60°N) and at 1400 m on the Yablonovoy and Stanovoy ranges at 57°N (Fig. 6).

The lower Lena valley has *tundra surface-water humic gleys* with permafrost at 50 cm. The highly acid humus is thick, and may extend to the frost table without any intervening mineral horizon. If present, mineral horizons are rich in sesquioxides as *As* horizons. Aerobism exists on gravel ridges, which remain *unpodzolized* because of the low rainfall (15 cm Ø).[40]

Throughout this extra-continental zone there is permafrost and a non-leaching climatic regime. Profiles are weakly developed, with thick organic layers; the soils are light brown in colour, show some gleying and much cryogenic mixing. Often there is an upward movement of soil solutions in autumn to the site of initial surface freezing, and dissolved minerals therefore concentrate at shallow depth. Thus *surface ferruginized gleys* are the main hydromorphic soil in the Tunguska-Vilyuy lowlands, while *frozen pale yellow sod-forest soils* and *slight solodization* are common. Further south (56°–50°N), *mountain tundra* occurs in the trans-Baykal, above 1600 m (Fig. 6); while below 1300 m *slightly podzolized shallow humic soils* are common under larch, grading into *mountain chernozem* near Chita (52°N).[2,27] In the borderland with Outer Mongolia coarse skeletal *chestnut* and *arid brown soils* occur.

Rainfall increases eastward into the upper Amur (53°N, 125°E) to 500 cm Ø where soils are more podzolized, with very acid *mountain taiga soils* (pH 3) and *sod-brown soils* (pH 5) on loams.

Far Eastern Siberia has an unusual pedologic layout.[26] Prolonged freezing gives *frozen ferruginous soils* and dry permafrost in mountains, *frozen humic soils* on piedmonts. Rainfalls are low and gleyzation is rare; lichens replace moss and grass, and soils are very acid. Podzolization occurs only on south-facing slopes in the more humid areas. The mountains of Verkhoyansk and Cherskiy[2] have *base-rich steppe soils* on warm dry slopes; permafrost and *frozen unpodzolized acid earths* on north slopes; *bald mountain sod soils* (=alpine turf?) on summits, and *taiga sod soils* on dry western slopes under larch. Thus far north-eastern Siberia has a 'sod' process, caused by dry frozen winters and warm, moist summers.

Oceanic Kamchatka, south of 58°N, has '*sod-birch*' *soils*, of brownish colour, and the Maritime Provinces have *forest brown soils*. Most distinctive are the soils of the cool monsoonal area south of the permafrost boundary—the white soils or *podbiel*[26] of the mid and lower Amur valley, the Khanka Lake, and the Sungari valley of N.E. Manchuria. Other names for this soil are *bleached meadow soil* and, in Chinese, *Peichiang-Tu*—soils 'white as milk'. They are 2–3 m deep, with an H, A2, AB, Btg, CG profile resembling *solodized sod-podzolic soils*, and show pseudogleying, for the soil is frozen in winter and moistened in summer, a distinctly non-podzolic regime. Similar white soils occur in degraded paddy soils of N. Honshu and Hokkaido and near Lake Tai in the lower Yangtze.[41]

Gray wooded soils occur in the mid-continental area of Canada in N. Saskatchewan and Alberta,[42] and in the Black Hills of N. Dakota and in Minnesota. They are found on base-rich parent materials in association with podzols on sands, gleys and peats. In Canada they are on very recent drifts, with 380 mm Ø rain and 0°CØ under spruce or larch forest; in the USA with 500 mm rain and 4·5°CØ. First noted in 1929 as 'grey bush' soils, the modal soil is now called *upland gray wooded podzolic soil*. They are thin soils (*c* 75 cm) with LF, A2, AB, B, Cca profiles. The A2 are platey structured, light grey and shallow (20 cm); the B are blocky, grey-brown and thick (50 cm); the C are thin, slightly altered parent material. The upper horizons are slightly acid (pH 5–6·5), despite a base saturation of *c* 50%, and 2:1 clays dominate.[42]

The gray wooded soils serve to illustrate the difficulties which arise from regarding 'podzol' both as a process and as a morphologic term. To quote Moss,[42] 'in recent years less attention has been paid to morphology . . . in some instances . . . podzols . . . appear to be defined solely on . . . chemical composition. The gray wooded are not identical to podzols in this respect, especially in their higher pH and base status and decreased accumulation of sesquioxides and colloidal humus, for much of the clay accumulated in their B . . . is . . . unaltered . . . and the structure is distinctive. Chemically gray wooded soils compare with gray-brown podzolics . . . but are morphologically very dissimilar'. Nor are the gray wooded identical to the Russian grey forest soils, despite their equivalent geographic position; for grey forest have nutty or prismatic structure; the gray wooded of Canada have platy A and massive blocky B horizons.

In Canada fine distinctions are drawn, too, between gleyed soils occurring in different parts of the zone of podzols. *Meadow gleys* develop under swamp grassland in association with brown and gray wooded soils. They have thick, dark organo-mineral horizons and mottled subsoils. They are base-rich, with calcic or saline subsoils, and only thin peat. *Dark gray gleisolic* soils develop under swamp forest, with thin dark organic horizons and neutral mottled subsoils. The latter are typical of the St Lawrence Lowlands and both have high potential as farmland when drained.

Eluviated gleysolic soils have acid humus over bleached mottled horizons, differing in colour and reaction according to parent material. Found in the lowlands with podzols and gray-brown podzolics, they are far less productive and are known as *slough podzols*.[42]

On well drained non-calcic materials in Canada, podzols and g/b podzolic soils dominate.[43] The latter are common in S. Ontario, associated with dark gray gleysolic, brown forest soils and peat. The southern parts of the Shield have shallow podzols and regosols on stony drift, lithosols and rock dominant areas, as well as acid gleys and peat. Some acid *brown podzolic soils* are found in the southernmost part, distinguished from gray-brown podzolics by the absence of free lime in the profile. In the podzol zone of the northern Appalachians, podzols, brown podzolics and eluviated gleys form the main associations.

1 S. Sjörs, Endeavour, XX, 80, Oct. 1961, 217–24

2 Y. N. Ivanova and N. N. Rozov, Sov. SS, 11, 1961, 1171–81 and TICSS, 7, 1960, V. 11, 77–87

3 P. Duchaufour, *Dynamics of Forest Soils under the Atlantic Climate*, Lectures given at the Faculty of Surveying, Quebec, Sept. 1958, 1–82

4 G. W. Dimbleby, Oxford Forestry Mem. 23, 1962, 1–120

5 V. V. Ponomareva, Poch., 3, 1956, 31–47

6 I. Barshad and L. A. Rojas-Cruz, SS, 70, 1950, 221–36

7 P. Duchaufour, Rev. For. Fr., 1951, 647–52; V. M. Fridland, Poch., 1, 1958, 27–38; B.C. Deb., JSS, 1, 1950, 112–22; L. O. Karpachevskii, Poch., 5, 1960, 43–52. A. A. Rode, *The Podzol-forming Process*, Akad. Nauk. SSR, Moscow, 1937. The mechanism of podzolization is reviewed in K. Kawaguchi and Y. Matsuo, TICSS, 7, 1960, V. 42, 305–13 and in the papers of C. Bloomfield:—

8 C. Bloomfield, A Study of Podzolization, JSS, 1953–5. I. Mobilization of iron and aluminium by Scots Pine, vol. 4, 1953, p. 5; II by Kauri, 4, p. 17; III by Rimu, 5, p. 39; IV by larch, 5, p. 46; V by aspen and ash, 5, p. 50. Also TICSS, 6, 1956, II. 5, 280–3

9 A. Muir, *The Podzol and Podzolic Soils*, Adv. Agron., 13, 1961, 1–56

10 G. F. Weetman and N. B. Nykvist, Pulp and Paper Res. Inst. Canada, Tech. Rept. 317, 1963, 1–15

11 Yu. D. Abaturov, Sov. SS, 6, 1961, 627–33

12 W. L. Kubiena, *Soils of Europe*, 1953, 257–66

13 N. Linnermark, *Podsol och Brunjord*, Lund Inst. Min. Pub. 75, 1960, vols. 1–2, see 1, 61–67

14 J. Låg, SS, 71, 1951, 125–7; also R. E. Wicklund et al., Can. J SS, 39, 1959, 222–34

15a B. T. Bunting, GJ, 130, 1, 1964, 506–12.

15b ibid TIBG, 26, 1959, 89–105, Fig. 2 and AJS, 259, 7, 503–18, Fig. 5

16 G. Scheys et al., TICSS, 5, 1954, IV, 274–81

17 W. H. Lyford, PSSSA, 16, 3, 1952, 231–5
18 E. Mückenhausen, 1960, profiles 34 and 40, gives a good impression of such soils
19 E. Mückenhausen, 1960, profile 43
20 E. Crompton, JSS, 3, 2, 1952, 277–89
21 W. L. Kubiena, *The Soils of Europe* (ref. 12)
22 O. Tamm, *Northern Coniferous Forest Soils*, Oxford, 1950
23 As far as possible the spellings used are those of the original authors
24 S. Pawluk, Can. JSS, 40, 1960, 1–14
25 See I. P. Gerassimov, Die Erde, 94, 1, 1963, 37–47
26 G. A. Liverovskii and L. P. Rubtsova, Poch., 4, 1959, 60–70
27 K. P. Gorshenin, *The Soils of Southern Siberia*, IPST, 1960
28 O. A. Polyntseva, *Soils of S.W. Kola Peninsula*, IPST, 1962
29 S. F. Tatarinov, Poch., 7, 1957, 13–21 and D. M. Rybotsov, Sov. SS, 7, 1960, 705–14
30 P. Kundler, ZPDB, 86, 1959, 16–36 and A. A. Korotkov, Poch., 9, 1960, 62–70
31 R. Tavernier, La Carte des Sols de l'Europe, Pédologie, X, 2, Gent, 1960, 324–47
32 O. V. Butuzova, Poch., 7, 1960, 87–95
33 ibid and G. I. Grigor'yev, Poch., 6, 1960, 53–65
34 S. S. Morozov et al., Sov. SS, 12, 1961, 1292–1300; also V. V. Gorbunov et al., Sov. SS, 11, 1961, 1182–7
35 E. V. Semina, Poch., 1, 1961, 29–39
36 B. P. Gradusov et al., Poch., 7, 1961, 59–66
37 I. P. Gerassimov et al., *Soils of Bulgaria*, Akad. Nauk., 1959, 1–398
38 A. V. Kolovskova, Poch., 3, 1960, 42–52
39 S. Pawluk, TICSS, 7, 1960, V. 43, 314–22
40 In Y. N. Ivanova et al., Sov. SS, 11, 1961, 1171–81
41 T. J. Yu, Acta Pedol. Sinica, 7, 1959, 42–58; also E. A. Kornblyum and B. A. Zimovets, Sov. SS, 6, 1961, 634–42
42 H. C. Moss et al., JSS, 6, 2, 1955, 293–311. On soils of the Black Hills see PSSSA, 27, 5, 1963, 573–6
43 P. C. Stobbe, 'The Great Soil Groups of Canada', in *A Look at Canadian Soils*, Agr. Inst. Rev., 2, 1960, 20–6
44 B. C. Mathews and R. W. Baril, ibid., 37–40

F

14

Brown podzolic and brown earth groups

THE concept and definition of 'Braunerde' was originally formulated by Ramann in 1905, and to it was added that of 'Podzolic Brown Earth' in 1908. Brown earths are widespread in central Europe upon a great variety of parent materials in a temperate climate, with moderate leaching under deciduous forest if in a natural state. The soil is almost uniformly brown, the colour and base status varying with parent materials and with organic matter content. The central part of the profile is a visually indistinct (B) horizon, sometimes slightly enriched with silicate-clay minerals or having some distinct development of structural forms. Normally this B horizon is altered by *in situ* 'weathering' rather than by illuviation. In contrast, *brown podzolic soils* have stronger leaching, and A2 horizons are lighter brown. They show some sesquioxide illuviation in thick, reddish or coloured, B horizons, with iron oxides coating the original soil material.

Of the available parent materials in central and lowland west Europe, most are on relatively young geomorphic surfaces, and parent materials vary from acid sands through calcareous loams to clays of various kinds, the clays situated on high lying sites, not necessarily in depressions, where, of course, gleys would result. All these parent materials develop through lithosolic or regosolic forms toward brown earths. The most acid and permeable only briefly retain their brown colour at the surface before assuming a bleached, podzolic A horizon and podzolization is rapidly achieved (Fig. 5, 20).

Less permeable, calcareous, materials—loess, marls and glauconitic or calcareous sandstones—leach less readily, have more biologic mixing, and slowly develop through a sequence of brown earths to *grey-brown podzolic soils*, for the clay content of their B horizons increases, which in itself lowers their permeability (Fig. 5, 13). Finally this Bt horizon is slowly destroyed and podzolic brown earths form. Such a sequence is known from parts of Europe and North America on older land surfaces, but in many parts it has been delayed by biologic mixing, by erosion, or by cultivation.

Certain distinctive parent materials—basaltic or ferruginous rocks —have either dark or reddish-brown soils. Similarly, many calcareous rocks, covered by loess or by brown-coloured soliflucted material in the recently periglacially-affected areas of north-central Europe, have *brown calcimorphic soils*.

Soils on elevated sites on clayey parent materials develop the mottling typical of gleization at some depth beneath a dark brown, granular or blocky structured surface horizon. Such soils have been termed *pseudogleys*, with A, Bg, CG profiles, the 'g' implying seasonal, rather than permanent, waterlogging.

There are many difficulties in matching up the terms for all the brown earths found in the various European countries and in America. Some national classification schemes depend on an assumed sequence of development from brown earths to podzols; others on the nature of the original parent material or the base status of the soil (the base status being the result of the original parent material and the degree of leaching).

Classifying brown earths by the degree of leaching and development of B horizons is more acceptable than classifying according to the base status, or the nature of the original parent material. Most acceptable is the approach of Duchaufour (1960) envisaging a sequence from 'sol brun' through 'sol brun[1] lessivé' to 'sol lessivé'— the last with a well developed Bt horizon.[,2,7] Further development produces an 'ochreous (yellowish-brown) podzolic soil' before the 'podzolized (pale grey) soil' is achieved.

We may now consider these soils in more detail. *Brown podzolic soils*, in the sense of 'soils with light or greyish brown mineral A horizons—completely devoid of carbonates and having lost iron oxides and clay—with coarsely mulliform surfaces and brown subsoils', are associated with cool temperate climates, acidic parent materials and with mixed or deciduous forest or, in parts of central Europe, with long cultivated areas. 'Podzolic' soils are of two very different kinds, either: (1) resulting from the slight bleaching of various newly-formed brown soils in sandy or loamy sand parent materials; or (2) the *parabraunderde* (=sol lessivé and the grey-brown podzolic soils?) on loess or fine sandy loams. In addition (3) there are the forms resulting from slight surface leaching of other groups—the *podzolic chernozem* and the *red/yellow podzolics*, for example—which are transitional soils of group status, with their own environment. The first two kinds are, by and large, stages between brown earths and podzols.

From these one may distinguish the *brown forest soil* of a loamy or fine texture, of high base status (base saturation >50%) and having intense biological activity. This is a more stable form, dependent on parent material for its base status, and resisting leaching because of its low permeability, water-stable structure, biologic mixing and return of base-rich plant ash to the surface. Such brown earths of central Europe are linked to a temperate or warm, subhumid climate in the oceanic-continental transition zone.[1] *In situ* weathering of soil components is intense, especially in the B horizon, but little translocation takes place and the upper horizons are enriched in free iron oxides and by the synthesis of clay minerals. As rainfall effectiveness is low (Table 8)—a summer maximum and high evapo-transpiration—leaching removes only soluble salts and some lime, and the other elements only slightly, if at all. Rapid decay of base-rich litter and intensive mixing by earthworms and other macro-organisms forms a deep mull layer. Iron oxides are flocculated by neutral organo-mineral colloids, giving the brown colour. Eventually weak lessivage occurs (the mechanical movement of unaltered clay to the B horizon) to form a textural B horizon or else, with time, in an aerated but moist B layer, more marked alteration of colour and some structural (blocky) forms develop.

Thus there are two subdivisions of brown earths dependent on the strength of the leaching process and the ease with which a distinct B horizon forms: I *brown forest soils* (=brown earths?) and II *sols* (*brun*) *lessivés* (=leached brown soils).

I. *Brown earths* have little visible differentiation of a B horizon for some considerable time and, apart from the surface mull, the soil is little removed from a deep, brown, ranker or regosol.[2] Clay contents are uniform with depth (Fig. 9, 1), the profile formula is (F), Am1 (or Ap), (B), BC, C. The A is crumb-structured, dark greyish brown and thin; its humus content varies from 3% to 10% and the pH is c 6; base saturation is c 50%. The (B) horizon is a neutral, brown horizon, 30–60 cm thick, followed by an equally indistinct C horizon, with pH c 7 on limey parent materials, yet lower on others. The free Fe content in the (B) may be slightly higher than that of the A or C horizons,[3] showing a 'colour B', and this horizon becomes clearer with time. Base-rich rocks, or those of low permeability but on well drained sites (schists on hills or marls on slopes) and well drained clay-loam materials, favour its fullest development.[4,5]

On base-deficient materials, and on well drained slopes on slates or shaley tills, *acid brown soils* (sol brun acide) (Fig. 5, 14), form,[1,6,7] which are easily podzolized to form *podzolic brown earths*. *Sol brun acide* would seem to belong to many parts of Hercynian Europe, along with podzols, on siliceous parent materials in humid areas. These coarser-textured materials are soon podzolized, rapidly developing colour B and then Bfe horizons at shallow depth.

II. *The sol brun lessivé* is characterized by downward translocation of clay size material, including ferric oxides, from a loamy surface to form a distinct brown, y/b, or more rarely r/y, blocky structured Bt horizon, with clay skins on ped surfaces and along channels. This is overlain by a slightly acid mull and brownish grey, A2–3 horizons, with a crumb, fine blocky, or else faintly platey structure. The C horizon is least acid and less fine than the Bt (see Fig. 9, 2). The soil has lower base status than brown forest soils though the Bt has >35% base status. The Am1 still derives bases from ash, if deep rooted vegetation is present, but when this is cleared leaching is enhanced.

The *sol (brun) lessivé* results from this increased leaching and corresponds to most soils termed *gray-brown podzolic* in America, and to the *sod-podzolic* of Russia, though this is best compared to the *sol lessivé* or German *fahlerde* (=buff earth),[8] which are more degraded, with very deep A2.[1,2,7] *Leached mull soil* would be a good English name for *sol (brun) lessivé*. They are common on the loess of Belgium and the Westphalia *Börde*, the Paris Basin and on the older Thames terraces; on brickearth and on other older surfaces in lowland Britain and on sandy loam-textured materials, formerly slightly calcareous. They are associated with *sols brun acide* on arenaceous materials and with *podzols* on coarse acid sands; also with *brown earths with gley subsoils* on moister argillaceous materials which have good surface drainage.[2,6]

The *grey-brown podzolics* of north-east USA are mainly found on sites north of the Wisconsin terminal moraine, on slightly acid parent materials of not too coarse a texture.[8] They have humid climates, with a summer rainfall maximum. Time has elapsed for them to be leached of Ca and Mg carbonates to 2 m depth and to have been oxidized as A/B horizons to 1 m.[9] On pre-Wisconsin land surfaces south of the moraine they are in a later stage of development with carbonates leached to 5 m depth, and are more deeply oxidized forming a transition to the red-yellow podzolics of southeast USA. In east-central USA (eg Indiana), gray-brown podzolics are found on parent materials younger than Illinoian till, on loess or on limey Wisconsin drift, developed under oak, hickory or mixed forest and now under corn and small grains (Fig. 14). They extend as far south as Louisiana on terraces and loess near to the Mississippi on materials *c* 10,000 years old. They are replaced by alluvial soils nearer the river and by red-yellow podzolics on yet older land surfaces. In the Appalachian area soils are usually podzols or sol brun acide, for grey-brown podzolics occur only on calcic material with thin plant debris layers; westwards they occur only on sandy materials among the brunizem and prairie soils. Fragipan, long cultivated, gleyed, solodic and planosolic variants are known, as well as *intergrades with brunizem*;[10] the change from brunizem to g/b podzolic taking place within 100 to 400 m, with greying of the A and more marked structural development in the B as the latter soil is approached (Fig. 5, 13, 16, 18).

Cline[8] considers the American *g/b podzolic* as a soil development midway between the *brown earth* (in the American sense of base-rich soil with a (B) horizon) and the *brown podzolic soil* (see Fig. 9, 3). In the brown earth decarbonation is not achieved, nor is an illuvial B. When bases are removed from the A and biological activity has decreased, the Bt of the gray-brown podzolic is formed. Later this Bt is destroyed in its upper part and the A horizons begin to show signs of podzolization (Fig. 9, 4).

Apparently the sandier the soil, the more readily the upper part of the Bt is destroyed, tending to move or be sited more deeply in the soil, the A2 deepening to >75 cm. While the plant roots are destroying the upper Bt its lower part is compacted, clay fills the ever-finer pores, impeding water movement. Reduction and pseudogleying occur,[12] and a braunerde-pseudogley develops (Mückenhausen, 1960, plate 38). The signs are ochreish spots and concretions in the lower part of the y/b A2, brought about by more frequent waterlogging above the deep-lying, thinned, but compact Bt (see Fig. 9, 3). The phenomenon is represented, too, by greyish or yellowish brown streaks along root channels and cracks, with reddish or yellow colours in the most compact parts of the Bt horizon.

The most usual names for such soils are pseudogley braunerde, or *sol ocreuse*.[8b] The organic layers are moder-like, and there is a slight development of a bleached A2, masked on cultivated land.

Later stages of soil formation, on somewhat coarser material (eg loamy sand), would give earlier destruction of a weaker Bt. Developed moder and deeper bleaching would be present. Such a soil is rather ambiguously called *podzolic brown earth*, with an indistinct B(s) layer which has apparently developed *within* the previously existing deep A2. They are termed *prepodzol* in Belgium, further degrading on fine sands[2] to a *semi-podzol*.[13] They are not to be confused with the *brown podzolic soils* developed on acid, coarser textured sands and medium textured sandstones which have shallower Ae, thick A3 and brown B2 horizons and no Bs; though, in a manner of speaking, the soil development on the finer materials has now 'caught up' with that on coarser materials.

Podzolic brown earths are found well developed on late glacial sandy tills, or on fine sandstones. Organic matter and base exchange capacity decrease regularly with depth, in striking contrast to the podzol.[10] The podzolic brown soils *in the alternative sense* of a very thin (2–10 cm) Ah or A2 (bleicherde) over a thin Bs and a thick, brown B, are a degradation of sol brun acide caused by bad forestry management (ie they are micropodzols). If a base-rich brown earth on a limey material (limey till or marl) suffers long continued leaching, under natural conditions, a grey-brown podzolic soil forms, followed by a podzolic brown earth with some Bt at the base. Afforestation of a clay or base-rich brown earth with, say, spruce, would not result in rapid podzolization and bleicherde formation; rather would it result in rapid growth of the trees.

Kubiena's approach[5] to the classification of brown earths (op. cit., pp. 230) is perhaps more convenient to the geographer than is the multiplicity of genetic sequences outlined, though the sequential approach is essential if the influence of time in soil degradation is to be understood. Kubiena separates out the brown earths mainly according to (1) geographic location, (2) parent material and (3) humus form—under natural conditions. He has two subdivisions of brown soils in 'humid and deciduous or mixed forest regions': (i) *climatic* and (ii) *a-climatic* brown earths; the first zonal (centro-European braunerde) on base-poor parent materials; the second, formed on base-rich rocks. There are many climatic or oligotrophic brown earths, and three forms of eutrophic brown earth.

Among the a-climatic forms which have been described in Britain is the *ferritic brown earth* which occurs on iron-rich parent materials —the middle Lias ironstone.[15] The soil is usually a 1 m deep clay loam, with slightly acid A1–2; neutral, dark, (B); and deep, coarser (B)/C, Cca and C horizons. The soils are rich in flocculated ferric hydroxides, there is no podzolization, and they are likened to sub-tropical red earths.

Other forms of brown earth are formed on siallitic parent materials which may overlie chalk. Such are the soils on the clay-with-flints and decalcified clays of south-east England and north-east France; they are of variable depth, texture and colour.[16] The soils are usually polygenetic, with yellow, brown or red colours at depth as residual rotlehm[5] or braunlehm, similar to Kubiena's 'plastosols'. They have become earthy at the surface because of alternating cold and warm conditions in the Pleistocene, additions of loess at the surface, and by disturbance and leaching under periglacial and later conditions. The resultant mixture in the upper layers is at present undergoing contemporary processes, with sol brun acide, pseudogleying or 'sol lessivé' according to site and texture. In some parts the old red and yellow soil materials are at shallow depth.

Brown forest soils do not occur widely in North America, only in the Lakes Peninsula, in New York, and in N.W. Washington. *Brown wooded soils* form a proto-gray wooded soil, having thinner profiles and moder humus to distinguish them from brown forest soils. Brown wooded soils form into podzolic soils, the brown forest into g/b podzolic and then into r/y podzolic soils.[17] Another Canadian brown earth is the *red-brown concretionary soil* on the southernmost coast of British Columbia. It resembles the subtropical soils of summer-dry areas,[18] and has an FH, Ah, Bfe$_{cr}$, Bt, C profile (Fig. 12b).

The brown soils of Mediterranean Europe are studied in Chapter 16.

1 R. Tavernier and G. D. Smith, *The Concept of Braunerde in Europe and the USA*, Adv. Agron., IX, 1957, 217–89; also I. P. Gerassimov, Poch., 7, 1959, 69–80; P. Duchaufour, *Précis de Pédologie*, Paris, 1960

2 J. Ameryckx, Pedologie, XI, 1960, 124–90; also H. J. Altemüller ZPDB, 98, 3, 1962, 247–58

3 P. Kundler, ZPDB, 78, 2/3, 1957, 209–32; also A. Thaer Ark., 6, 2, 1962, 111–17 and ZPDB, 96, 1, 1962, 62–70

4 E. Ehwald, ZPDB, 80, 1, 1958, 18–42

5 W. L. Kubiena, *Soils of Europe*, 1953, 230–42

6 D. Mackney, JSS, 12, 1, 1961, 23–40

7 B. W. Avery, JSS, 9, 2, 1958, 210–24 and TICSS, 6, 1956, V. 45, 279–95. Also B. W. Avery et al., JSS, 10, 2, 1959, 177–95

8 M. G. Cline, SS, 68, 1949, 259–72; and PSSSA, 18, 1954, 148–53. R. Tavernier and R. Maréchal, Pedologie, VIII, 1958, 133–82

9 E. Frei and M. G. Cline, SS, 68, 1949, 333–43; and with S. B. McCaleb, SS, 77, 1954, 319–33

10 H. C. Wascher et al., Illinois Agric. Expt. Stn. Bull. 665, 1960

11 USDA, *Soil Survey of Brown Co. Kansas*, Ser. 1957, 7, 1960

12 P. Duchaufour, *Pédologie, Applications Forestières et Agricoles*, Nancy, 1956, p. 192

13 Term suggested by J. Fink, Mitt. Österreich Bodenkundl. Ges. 4, 1960, 45–58; and in *Die Böden Osterreich*, Mitt. Geog. Ges. Wien, 100, III, 1958, 92–134. See also ref. 6

14 Kubiena applies the term 'Semipodsol' in a wider sense, corresponding to the wider or first sense of 'podzolic'.

15 R. Storrier and A. Muir, JSS, 13, 2, 1962, 259–70

16 B. W. Avery et al., JSS, 10, 2, 1959, 177–95

17 P. C. Stobbe, PSSSA, 16, 1952, 81–4

18 J. S. Clark et al., SS, 95, 5, 1963, 344–52

Soils of subhumid plainlands

MOST grassland soils of central North America and Eurasia are cultivated and are *chernozemic*. They have blackish, deep, mineral-organic A horizons; thin, dark brown B horizons with clay accumulation; and calcareous subsoils. They are mainly base-saturated and salt free; more so in the humid eastern or northern parts where the carbonatic layers lie deepest. With increasing leaching (from west to east in the USA, S.E. to N.W. in Eurasia) the sequence of soils is—calcic brown; chestnut; black (chernozem); leached and podzolized chernozem; brunizem and prairie soils, the prairie soils intergrading to gray-brown podzolic soils.[1,2]

Prairie soils occupy much of Iowa, upland Illinois, and eastern parts of Kansas, Oklahoma and Texas. They are replaced by gray wooded soils in Canada. Their latitudinal extent implies variations due to climate and to changes of parent material: Pleistocene deposits in the north, Tertiary rocks in the south.

'Prairie soil' has a restricted meaning in soil science, relating to complete decarbonation without movement of sesquioxides. In Illinois on loess, they are inceptisols, often planosolic, varying with the texture, depth and lime content of the parent loess.[3] In eastern Illinois they are degraded or even podzolized by tree invasion, which has produced light A2 horizons in 100 years, transforming degraded prairie soils into gray-brown podzolics.[4] The more markedly summer-dry prairies of Iowa and Kansas, which are more continuously frozen in winter, are less leached. This is the area of modal prairie soil, with vestiges of carbonates at 1–2 m.[5]

Brunizem (Fig. 5, 18) as a modal form are sited on young (late Wisconsin) land surfaces in humid areas (Fig. 15). Rainfall leaches carbonates to give flecks in a C2ca at 1·5–2 m depth if on calcic loess; to limey coatings on pebbles at 0·5 m if on compact till. The profile formula is Ah1–3, (AB), B2–4, C1–2, C3ca; the A horizons being friable, dark g/b, clay loams, <50 cm deep; the B, >50 cm thick, blocky and yellowish brown; the C extending from 70 to 170 cm and being yellowish, silty, with lime at the base. There is no Bt horizon, and organic matter is found as deep as the B3 horizon. The brunizem is a distinct group long since deforested by Indians, bison or both; and the possibility of reafforestation is remote, for they form highly productive land.[4] A profile similar to brunizem has been described in Germany, near Grossenbröde, on Würm 3 (Pomeranian) calc-till.[5]

Reddish prairie soils, more altered, occur on older land surfaces to the south, or on old Kansan till in the north though on surfaces younger in date, cut into the till. Erosion of A layers, or mottling of C horizons by a high water table, are common on these northern soils. Modal reddish prairie soils occur in Texas on clayey alkaline parent materials, while sands have thin red podzolic soils. Reddish prairie are highly productive of both crops and grass; the upper horizons are slightly acid, granular, clay loams, humus-rich, and grey-brown; the B are reddish granular clays; the C, red or yellowish brown blocky horizons with shot-like Fe and Mn concretions.

The *chernozem* proper—calcaltoll 7A—are associated with long-grass steppe; a continental, subhumid, climate with warm, moist early summers; late summer and autumn drought; and dry, relatively snow-free winters. Little leaching occurs in the frozen soil in winter, and the moist early summer is counteracted by intense evapotranspiration. Rapid growth of grass or spring-sown cereals occurs in a moist soil in early summer, and rainwater rarely penetrates to the base of the profile; if it does, it only removes soluble salts into the drainage waters. Grass growth is limited by late summer drought; leaves and fine roots die off, and humify in the following year.

Many lime-rich parent materials produce chernozem—loess, basalts, diorites, gabbros, calcic or fine grained sandstones, marls and terrace alluvia. Coarse sandy materials in the chernozemic zone give rise to 'podzolic' soils. The *modal chernozem* has a dark surface horizon, 60 cm deep in the USA, 90–150 cm in Russia, with 5%–20% organic matter by weight of the soil, 50%–65% of which is organic carbon, dominated by humic acids. Usually the Ca-humic acid combination immobilizes this large mass of humus; though the upper 20 cm may be lighter in colour, have pH 6·5, and granular structure as an Ah1 or Ap. Beneath are thick, darker, prismatic-structured A21, with pH 7·5 and a less humic A22. An AB layer is greyish brown, shallow and irregular and has a low organic content (Fig. 11). The inner parts of peds are here markedly limey, for leaching is only along channels. The C horizons have slightly less clay, but are rich in lime (c 15%) and much lighter in colour. The basal Cca is at 90 cm in the USA and is highly calcic (20%–35%).

The depth of Cca is related to precipitation: with 50 cm rain it is at 50 cm depth, with 75 cm rain at 1–1·2 m (E. Nebraska) and with 1 m rain at 1·5 m depth. Yet these are general figures for a west-east gradient, and many other factors affect its depth. It is at shallow depth on flat sites or in uplands, at greater depth with increasing depth of ground water or with lighter textured soils. The depth of solum above the Cca increases downslope, it is deepest in well drained concave depressions, and may be absent from periodically flooded or waterlogged sites. It is friable, with large, weak, structural units, and exists as closely-spaced flecks, spots, blotches, 'white eyes' (infilled burrows), or concretions, rather than as a continuous layer.[7] It has pH >8 and a Csa occurs beneath it if subsoil drainage is poor (see also Fig. 3).

In the USA chernozem occur in the eastern Dakotas, S.E. Nebraska and in north-central Kansas. South of 37°N are *southern reddish chernozem* with lighter brown surfaces and yellow or red subsoils, distinguishable from the reddish chestnut soil by darker surfaces and by lower chromas in the subsoil. One may also distinguish eastern and western forms of chernozem in the USA; the eastern are more moist and darker, with A1p, A12, B2, B3, Cca, C profiles; the drier western *greyish brown chernozem* have only 50 cm rain and an Ap, B2, B3 (ca), Cca, C (sa), C profile.

In Canada, chernozem occur in S. Manitoba, in central Saskatchewan and in Alberta, as far north as the Peace River at 57°N.[2] Grassland and 'park' areas have *black chernozem*, with more humus than in the USA. If eroded, the blocky, fine textured B is at the surface. Low-lying areas have *solonetsic soils* reflecting the early stages of weathering on Cochrane and Valders tills, or on Lake Agassiz deposits. Thus the American (and Eurasian) chernozem may be explained by morphology (profile), and by geographical distribution. All show intense, deep-reaching faunal activity.

In Eurasia chernozem are widespread.[8] The westernmost area of *modal chernozem* is south of Wroclaw, but *degraded chernozem* occur as far west as Braunschweig. South-east Europe has more typical chernozemic soils—in W. Slovakia, central Hungary, N. Bulgaria and S. Rumania. The Russian 'belt' commences in the Pontic–S. Bessarabian area (28°E, 46°N) and trends E.N.E. to the Urals (at 55°E and N); the Ural mountains cause their displacement southwards. The chernozem then extend due east to 83°E, meeting the Altai Range between 52° and 55°N. They recur in parts of E. Mongolia and N. Manchuria.

The criteria for subdivision of Russian chernozem are (i) organic matter content; (ii) relative depth of A-layers and (iii) depth of effervescence—the level at which free carbonates are found. The sub-types of chernozem are six in number:[1,9,10]

1. *Degraded chernozem*, in which humus is moved from an Ah1 to an Ah2, and they may be leached or podzolized. (1a) *Podzolized chernozem* occur under forest, have distinct LF layers, and 50 cm deep, quartz-rich, slightly acid A horizons. The upper Alh has 5% humus; the thicker A2h has 8%. At 1 m depth there is more humus (10%) and reddish B(s) are visible. The Cca lies very deep (c 2 m). (1b) *Leached chernozem* form under grass, have 7%–12% humus in the 50–80 cm thick, neutral A layers. The soils occur in the most humid parts of the Russian chernozemic zone, and differ from the grey forest soils in their darker colour, neutral reaction, more soil fauna, and in having a calcic layer at depth and deeper distribution of humus. Chernozems in central Europe are degraded, with reddish (B), and are located on loess or late-Tertiary marls, as in the Vienna basin[11]—the *Danubian chernozem* having mycellial carbonate layers at c 1·4 m and light yellow subsoils. They are termed 'brown chernozem' if on older terraces. The brownish chernozem of the northern Balkans is transitional to the meridional brown earth, the degraded Russian forms to the grey forest soil.

2. *Typical chernozem* is the modal form in moist areas, with long grass, varying in humus content from 5% (Ukraine) to 10% in Trans-Volga. It is the most humus-rich form (>500 tons/ha/m). Depth of effervescence is 70 cm in the Ukraine—where the full

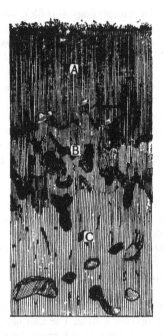

Fig. 11 Dokuchayev's illustration of the chernozem. Presumably a typical deep form. Note the irregular 'pocketed' (B) and *crotovina* in the C

depth of organic layers is 1·2 m, or at the base of the organic layers in trans-Volga (60 cm depth). Though such thick and thin chernozems are related to climate, relief has similar effects. Thick chernozem form on level sites; thin on slopes >6°, or on compact parent materials. Thin chernozem have lower contents of organic matter, no (B) horizons, and are droughty.

3. *Ordinary chernozem* occur in slightly drier areas under grass of medium height.[12] Humus contents vary from 6% to 10% and the A horizons are 90 cm deep in the Ukraine, 50 cm in N. Kazakh. Humus contents average 400 t/ha/m. Effervescence occurs at *c* 40 cm, and gypseous concretions may occur at 2 m depth, marking the limit of moisture penetration.

4. *Southern chernozem* form under short grass in the drier parts of the chernozemic zone. They are dark grey in colour, and have 3%–6% humus, the powdery A layers being 40–70 cm deep. Mg, rather than Ca, is the dominant cation and readily soluble salts are found at 1·5 m. 'White eyes' are very common and depth of effervescence is only 20 cm. Apparently humus is slightly leached by mechanical eluviation, not in solution, as happens further north.

5. *Carbonate chernozems* are more localized, with shallow (40–80 cm) A layers, effervescing at the surface, and with 300 t/ha/m of humus. Residual forms occur on limestone and resemble rendzinas; *mycelial chernozem* are found in the warm, moist Azov-Ciscaucasian area, and have thick, dark brown nutty-structured A layers, containing needles of calcite.

6. Clay-rich *compact chernozem* have little decarbonation because of their high content of physical clay (70% <0·01 mmø in the Russian textural classification) and 50% true clay (<0·001 mmø). They effervesce at 25 cm, are rich in humus (550 t/ha/m), with a 1 m thick blue-black A, and a very compact Bt1 with 60% true clay. The soil is markedly hydrophilic, swelling on wetting and with ferrous iron near the surface. A typical parent material is, for example, Tertiary clay, as in central Moldavia.

Other forms of chernozem are chestnut, meadow (warp), and deep effervescent chernozem—the last a transition from typical to leached forms, common in E. Rumania. A chernozem much affected by earthworms is known as a vermudoll (7A), and other forms are gleyed, solonetzic (near Lake Chany), solodic, mountain and solonchak-chernozem.

Reddish chestnut and chestnut soils are the next group in decreasing order of rainfall effectiveness.[13] The American and Eurasian zones both have 35–40 cm rain, short grass vegetation, and alkaline organic matter. Horizonation is indistinct; the surface horizon a grey-brown, friable, platey structured Ah/A2, grading through an A3/B1 to a columnar B2ca. Clay accumulation and gypseous specks are marked in a compact B3. The calcic horizon is much nearer the surface than in chernozem, though the cause is more complex than decreased rainfall alone. The slower decay of a smaller supply of plant debris produces less CO_2 and less intense solution of $CaCO_3$. CO_2 pressure of soil air also decreases more rapidly with depth than in cooler, more humid, areas, and so $CaCO_3$ is more readily and shallowly precipitated.

In N. America, *chestnut soils* are best developed in western N. Dakota and S. Saskatchewan. At greater altitude in N. Montana a *dark chestnut soil* forms because there is slower oxidation of a closer grass cover. In Dakota the soils are more grey at the surface. Fine textured soils under Buffalo grass are productive, because drought resistant. Under Bluestem grass on sands they are greyish and non-calcic at the surface and easily erodible because of poor structure.

To the south, in W. Oklahoma and N. Texas, a *reddish chestnut soil* is formed, with an A1, B1, B2, B3ca, Cca, C profile. On calcic or clay-rich materials, the B2 is clay-enriched and the Cca is formed of pinkish platey *caliche* at 1 m depth, with 20%–50% lime content, varying in thickness from lime coatings on pebbles to indurated horizons several feet thick. It may also be a relic feature, overlain by wind-blown sand as in the deserts of New Mexico. The chestnut soils vary widely in texture, depth, colour and in depth of decalcification, and the subsoil colour is greatly influenced by that of the parent rock.

The chestnut soils of the USSR form a narrow belt occupying the dry steppe. The western part comprises the coastland of the Black Sea from Odessa to Karkinitskiy, and those of the northern Crimea and Azov. The western Ciscaucasus area intervenes with its chernozemic and mountain brown- and grey-forest soils, before chestnut and solonetsic soils recur at 42°E. The soil zone here significantly turns northwards from 44° to 51°N, coinciding with the former extent of the Caspian Sea,[14] before turning east to form a wide zone between 47° and 52°N from the Volga at 43°E to Sinkiang at 84°E (Dzungaria). Chestnut soils thus occupy much of *Kazakh S.S.R.*, with 75% of the area as *chestnut* and the rest as *solonetsic* soils.[14]

In this longitudinal stretch, climate differs markedly, with warm, moist conditions in the west (400 mmØ), decreasing to 350 mm in the north-east and 180 mmØ in the south-east. In the west the January temperature is −1°C, in the east −18°C. Summer temperatures are the same throughout the zone (21°–24°C), and the vegetation is short *artemisia* grass of variable density.

The Russian classification of chestnut soils is of dark, typical and light forms. *Dark chestnut soils* have 0·5m thick A layers with 5% humus; *typical forms* have 3% humus; *light soils* have 0·3 m thick A layers and 2% humus, and are usually solonetsic, with Na accumulation at 0·2 m, peptizing soil clays and consolidating the base of the organic horizon. The solonets here is therefore an hydromorphic variant of a zonal chestnut soil, with a shallow water table causing capillary currents in summer.

Chestnut soils are also found in the Santa Cruz province of Argentina and in the Great Karroo of the South African Republic. Also in Tsinghai, Shansi (light) and Kansu (typical), on rather coarse loess.

The calcic or arid brown soils of semi-desert areas occur with sparse bunch grass or scrub and a rainfall of <35 cm; with the effectiveness of the rainfall reduced by high intensities and irregular incidence as summer storms.

In the USA these soils occur in S. Montana, E. Wyoming, E. Colorado, and in the upper Colorado area, and the Snake-Bear River Plains of S.E. Idaho. Calcic brown soils also occur in central Turkey, in the central Asiatic Republics, in W. Manchuria and in E. Mongolia. They are all mainly sandy, and saline, with Na salts and gypsum if there is a shallow water table or low rainfall (c 10 cm). They are *light brown* or greyish brown in colour, and surface humus and carbonate contents are <2% in the A1; hence the upper 10 cm of soil is leached, while the rest of the profile is alkaline. The A2 is nutty structured, the A3 is vermiform. The B horizon is clay-enriched, prismatic in structure and has calcic veins or spots. The Cca is at 30–40 cm depth and is saline—Csa—at 40 cm.

Reddish brown desert soils occur in Coahuila and in western Texas on sands with deeper (<1·5 m) A1, A2, B2, C profiles. They are neutral in reaction with single-grained grey-brown A and porous, red, non-calcic subsoils. They are also found in the eastern loess region of Shensi, with a similar hydrothermal regime.

Other semi-desert or desert-steppe soils are greyish brown soils (*gypseous serozems*); and grey desert margin soils (*true serozems*) with ephemeral therophytic vegetation and 1 m deep profiles. All these semi-desert soils are potentially highly productive on irrigation, for their base-content is high and they have developed on alluvial fans and playas, sites which are readily irrigated.[16] The fundamental contrast between grey serozemic soils of cooler arid areas and the red forms of subtropical deserts has already been drawn (p. 142).

1 E. N. Ivanova and N. N. Rozov, Poch., 12, 1958, 48–59; 1, 1959, 59–70

2 W. E. Bowser, in Agr. Inst. Rev., 1960, 15, 2, 24–28

3 G. D. Smith et al., *Adv. Agron.*, 2, 1950, 157–205

4 H. L. Wascher et al., *Characteristics of Soils Associated with Glacial Tills in N.E. Illinois*, Univ. Ill. Agr. Expt. Stn. Bull. 665, 1960

5 J. K. Ableiter et al., *Soils of the north-central region of USA*, Agr. Expt. Stn. Madison, Pub. 760, 1960

6 E. Mückenausen, 1960, profile 11

7 A. Bascó, Agr. Egyet. Mező. Kar. Közl., 1960, 327–38. Also J. E. McLelland et al., PSSSA, 23, 1959, 51–6

8 V. V. Dokuchayev, *Ruskii Chernozem*, St Petersburg, 1883, in Collected Works, vol. 3, Moscow, 1949, Also P. Kostichev, *Chernozem of Russia*, do. 1886. (Selected Writings, Moscow, 1951)

9 M. M. Kononova, in *Soil Organic Matter*, Pergamon, 1961, p. 361–4

10 See ref. 1 and A. A. Rode, *Soil Science*, IPST, Table 113, on page 477. On leached chernozem see A. V. Koloskova, Sov. SS, 8, 1961, 871–80

11 H. Franz, Bodenkultur, 8, 2, Vienna, 1955

12 I. P. Sukharev, Sov. SS, 2, 1961, 142–50

13 R. Smith, SS, 67, 1949, 209–18

14 G. G. Yeremin, Doklady (Pedology), 4, 1961, 74–9

16

Soils of Mediterranean
and humid subtropical areas

GEOGRAPHERS have perhaps overemphasized the distinctiveness of this climatic type, but its effect in soil formation has been little recognized by pedologists except by Duchaufour,[1] for whom the only unifying climatic factor is summer drought, though even this varies in duration and intensity, as well as regularity. Further climatic variables are those of latitude, altitude, aspect, oceanity and humidity, all affecting the soil.

Four main variants of Mediterranean soils, related to length of dry season and general aridity, may be recognized: (1) the *humid subtype*, with a short summer drought, well marked in mountain areas—the Tell and middle Atlas, S. Andalusia, W. Yugoslavia, Albania and central Chile—all have *leached red and brown soils*; (2) the *subhumid* areas, with >6 weeks' drought, have mainly *red soils*—Provence, the Catalan and Languedoc littorals, Corsica, Tuscany and western Greece; (3) the *semi-arid*, with 12 weeks' drought—the Algerian littoral, Sardinia, S. Italy, Sicily, which have isohumic *brown calcisols* and (4) the arid subtype, with 22 weeks' drought, in N. Africa (east of N.E. Tunisia), and the Almerian and Murcian coasts of Spain, which have *brown soils with calcic crusts*.[2]

Intrazonal soils also occur; *rendzinic soils* on soft calcareous marls, often eroded; and the black *Tirs*—black or dark brown vertisols, sometimes with silcrete. Also halomorphic and aridisols occur on desert margins, with calcic or gypsic dune soils. Parent rock and relief are potent factors in the Mediterranean area.

The first two variants—the red and brown soils—vary in age and depth, differing from those of temperate regions in their higher pH, more compact structure, and lack of organic matter[3,4] (Fig. 12a, and Fig. 5, 4, 6). Illite usually dominates in these soils,

1. *Terra fusca* (brown soils) are neutral or slightly alkaline, and are associated with *cork oak* in humid areas, or with *maquis* and moist site climates on siliceous rocks. They can form by degradation of terra rossa, with brown A over relic red Bt horizons. True terra fusca have a thin grey mull, with earthworms; compact, clay-rich, weakly acid Bt with ferri-hydroxide skins on ped surfaces formed during dry periods; and a C of dry carbonatic loam with y/b encrustations. *Terra fusca lessivé* are also found on the Tell and Kabylie and on the high mountains of Lebanon.

2. *Terra rossa* may form from t. fusca after deforestation and exposure to intense insolation, but mainly occur on calcic or dolomitic materials along with *garrique* vegetation. They are clay-rich red soils, full of rock debris and of variable depth—deep in fissures, very shallow on compact blocks of limestone (Fig. 12). Terra rossa are native to the littoral and continental karst[5] of S.E. Europe—Austria, Dalmatia, S. Italy and Sicily, Greece and Crete—the limestone lands of central Spain and the hills of north and central Israel. In the absence of a plant cover, or after deforestation, they are severely eroded,[6] though climatic rehydration causes progressive loss of free iron and formation of terra fusca and strong lessivage of clay under the influence of colloidal silica.

Formation of terra rossa is linked to the process of *rubification*, which is also typical of tropical ferruginous soils. Reddening by free Fe oxides occurs on decarbonation, by leaching in the wet seasons and reddening in dry periods; the ferric hydroxides forming electro-negative ferri-silicic complexes which are destroyed in the dry season, and hematite, goethite, or turgite are precipitated.[8]

[7] Degree of lessivage and depth of Fe reprecipitation vary with length of dry season and intensity of rainfall. Leached red soils have more humus and form in wet areas and show a Btfe horizon, more than usually enriched with iron. Often summer dewfall helps keep the surface moist and leached, as in the 'Atlantic' parts of the zone, or in mountainous areas.[3]

The parent material of terra rossa may be of a palaeogeographic nature, resulting from weathering in a moist subtropical climate in an interglacial period—proved by the kaolinitic clays in some soils

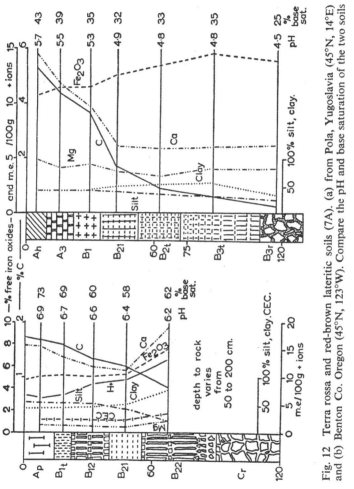

Fig. 12 Terra rossa and red-brown lateritic soils (7A). (a) from Pola, Yugoslavia (45°N, 14°E) and (b) Benton Co. Oregon (45°N, 123°W). Compare the pH and base saturation of the two soils

of southern Spain. Yet most reports show that, though parent materials are old, the present terra rossa are not in disharmony with the present climate. Under subhumid conditions *rendzinas* commonly develop into *terra rossa*; in alpine or humid areas they form *brown rendzina* and then *terra fusca*. Some Yugoslav workers consider that rendzinas occur with rapid weathering; with slower weathering terra rossa result.[5]

In arid areas the *crusted brown soils* of the *alfa* grasslands are found—in Europe, in Murcia and Almeria. In Morocco such soils have a brown carbonatic A, *c* 25 cm deep, termed *trab*; a subsurface calcic crust Bca(r), 10 cm thick and flakey, termed *tifkert*; and a tuff-like C horizon (*tafezza*).[9] These are common in the Moulouya valley, the crust forming after deforestation when soil water is submitted to strong evaporation. Strong sheet erosion by the winter rains exposes the crust.

Various perhumid types of *terra rossa* are known. In southern Portugal, with high humidities and rainfall (2160 mm), red-yellow podzolic soils form, associated with relic grey-brown podzolic soils.[11] Terra rossa are not to be confused with the *terra roxa* of São Paulo, which are a base-rich, low-humic, clayey latosol.

An important relic soil is the *relic rotlehm*, widespread in the high plains of the Guadiana, with lateritic concretions formed under subtropical conditions on Pliocene sediments. *Red weathering crusts* of lateritic type are also known from Albania,[12] with *leached cinnamon-brown soils* developed on and around them.

Such *cinnamon-brown soils*[13] are recognized in southern Bulgaria as a transitional form from the chernozemic or meridional brown soils to the Mediterranean red-brown soils, on hilly terrain.

Soils of the moist subtropics
Somewhat apart from the soils of the Mediterranean area are the *red/yellow podzolic soils*[14]—the Abkhaz littoral of the eastern Black Sea in western Georgia, presumably extending into the northern Turkish coastlands, and the Lenkoran littoral of the S.W. Caspian. The Lenkoran littoral has a drier summer; the Abkhaz littoral, perhumid summers, though without rain.

The Black Sea littoral is protected from the north by the Caucasus and has 2 mØ rain; it has natural, but now largely cleared, forests of beech, rhododendron and laurel, with bamboo on wet sites. In the north (Sochi-Abkhaz) the air is saturated in summer and yellow *zheltozem* occur, having iron concretions and gleyed C horizons on clay. To the south red *krasnozem* form on conglomerates, andesites and basalt.[15] They resemble the rotlehm (red loams) of Kubiena, dominated by iron hydroxides in both B and C horizons. Cloddy Ah horizons with 1%–2% humus and pH 4·5, overlie bleached grey-brown A2 and red-brown, prismatic, B horizons.

These Russian soils are very similar to those of S.E. United States and S.E. China (south of 33°N)[14], for both podzolization and laterization are active in the soil and in the weathering crust respectively. Colloidal Si is leached into the drainage waters or absorbed by bamboo, while liberated sesquioxides are kept near the surface by the vigorous vegetation, especially by the tea plant, and are not moved by cheluviation. Gibbsite is present in these soils.

The *red/yellow podzolic soils of south-east USA* are deeply weathered, shallowly-podzolized soils, often much eroded. They occur on many different parent materials and geomorphic sites.[16] They are usually well drained, and are acid, devoid of free lime, and have thin FH layers. Moistened Ah horizons may be black or chocolate coloured, becoming dark reddish-brown on drying, with crumb structure. Leached A2–3 horizons have less humus and heavier textures (Fig. 5, 12). The lower boundary of the A layers is indistinct, passing into a bright coloured B horizon, which may be red, yellowish-red or yellow, clay-rich yet friable, permeable, and firmly structured. The B are kaolinitic, and rich in free Fe and Al oxides. A B1 has few clay skins, but the deep B2t is brighter, has large peds and pronounced clay skins. The B3 is slightly mottled and clay forms *in situ*. The B3 and/or CR horizons have lower clay contents and often show the structure and colour of the original rock (see Fig. 5, 12).

The group has both red and yellow members; the *red podzolics* have thin y/b A2 and illuvial red B horizons; the *yellow podzolics* show thicker, yellow A2, and deeper, yellow B horizons.[17] The reasons for this colour difference are not entirely clear. Generally— with exceptions—*yellow podzolics* form on the more acid parent materials and are internally less well drained. It is also possible that they have formed under pine rather than under deciduous forest, or under a forest with little or no ground vegetation. Cherty material and fine sandy residues give yellowish-brown A and light yellowish-red B horizons. Coarse sands give vivid red colours with highly acid, bleached, A horizons, while truncated soils tend to be reddish. The Piedmont with its crystalline rocks, and the S. Appalachians with its hard pre-Carboniferous rocks show *red podzolic soils.* Some broad high-level peneplains have yellow soils, and the Coast Plain is also dominated by hydrated yellow soils.

Red and yellow podzolics have higher contents of Al and Fe than northern podzolized soils (Fig. 3), but much less Fe than true tropical ferrallitic soils. Al is present in silicate rather than hydroxide form, though silicate minerals mainly occur at depth. It will be recalled that silicate minerals accumulate at shallow depth in the Bt of grey-brown podzolic soils.

Several parts of the world have similar quasi-monsoonal climates and red and yellow podzolic soils. In their cooler parts, with some winter rainfall, the degree of surface podzolization increases, while in their warmer, equatorward, parts, ferrallitic tendencies are more marked, especially on well drained sites. This is the case in Georgia (USA) at 33°N, while, in Cuba (22°N), full 'latosol' development is evident. Red and yellow podzolics and red ferritic soils are associated in South America between 20° and 28°S; in eastern China and Szechwan (the Red Basin), between 23° and 32°N.[14] Both forms of red soil are found together in south-east Australia, while Tasmania has *krasnozemic soils* (22°–42°S).

A soil transitional from r/y podzolic to latosolic is the *reddish-brown lateritic soil* of Georgia developed on limestone or basic metamorphic rocks[18] (Fig. 12, 2). Lack of organic matter and of bleaching in the dark, red-brown A2 distinguish this soil from the r/y podzolics, as well as its greater depth (5 m) and higher clay content (80%) in the B22. Soils in the southernmost Piedmont are sticky when wet and crack when dry. Colloid content is high, kaolinite dominates, and free Fe increases from 2%–5% in the A, to 9%–12% in the B3 horizon. The clay is of coarse calibre (2 to 0·2 u) and silt is present in appreciable amount (Fig. 12, 2), unlike ferrallitic soils which have little silt and have clay of medium calibre (see also Fig. 3).

The Russian interpretation of humid subtropical soils is somewhat different. Afanasiev (1931)[19] and later Fridland (1961)[20] claimed that the zonal soil is a yellow soil, the red being an earlier stage, or else native to iron-rich materials. In many coastal lowlands of the subtropics (N. Vietnam, 20°N), bamboo forest is found on allitic yellow soils, under anærobic conditions. In E. Yunnan[21] (105°E, 23°N), the lowland monsoon-evergreen forest has yellow soils, while bamboo forest, a secondary growth after cutting and burning of virgin forest, has red soils. The hydrophilous nature of subtropical clay soils leads Zonn[21] to challenge the idea of a leaching regime in subtropical soils, except for the surface of red soils. Though bases and SiO_2 are quickly removed in the early weathering stages, rubefaction proceeds and is stable for long periods.

Thus subtropical soils consist of red soils, superficially podzolized, adjacent to, or intermingled with, yellow soils which are either on older land surfaces or in moister sites, and are more nearly zonal. In the northern hemisphere subtropical soils range from 46°N in Lombardy and 48°N in Washington State to 22°N in Yunnan. In the southern hemisphere they range from 21°S in Sao Paulo to 42°S in Tasmania.[22] They vary locally, especially in colour and texture, and on a global scale they vary according to a monsoonal, a Mediterranean, or a quasi-perennial moist regime—as in Yunnan and Russian Georgia.

1 P. Duchaufour, *Précis de Pédologie*, Masson, Paris, 1960. Also see J. Boulaine, *Sols Afr.*, VI, 1961, 263–82

2 See ref. 33, chapter 9

3 J. M. Albareda and A. O. de Castro, *Edafologia*, 1955, 2nd ed.

4 There are many studies of the range from humid to dry Mediterranean soils; many referring to the soils of one nation—A. Muir, JSS, 2, 1951, 164–82 (Syria); A. Reifenberg, *The Soils of Palestine*, Murby, 2nd ed., 1947, and in JSS, 3, 1952, 68–88. D. Yaalon, Bull. Res. Cncl. Israel, 8G, 1959, 91–118, and J. Dan and H. Koyumdjisky, JSS, 14, 3, 1963 12–20 (Israel). O. Strebel, Geol. Jb., 84, 1965, 1–22

5 M. Čirić, Zborn. Poljopr. Fak., Univ. Beograd, 7, 277, 1959, 1–12

6 I. Seginer et al., Bull. Int. Ass. Sci. Hydrol., VII, 4, 1962, 79–92

7 D. L. Bramao, *Carta dos Solos de Portugal*, Lisbon, 1949

8 P. A. Anastassiades, SS, 67, 1949, 347–62

9 J. Boulaine, CRAS, Paris, 253, 1961, 2568–70

10 D. L. Bramao and R. W. Simonson, TICSS, 6, 1956, V. 4, 25–30

11 J. V. J. de Carvalho Cardosa, TICSS, 7, 1960, V. 9, 63–70

G

12 K. P. Bogatyrev, Sov. SS, 10, 1960, 1054–61

13 I. P. Gerassimov, Izd. Ak. Nauk. SSSR, 1954 and G. Gyurov and N. Ninov, in *Soils of S.E. Europe*; Acad. Agric. Sci., Sofia, 1964, 113–32

14 B. B. Polynov, CRICSS, 5, A.1, Soviet Section, 1935. On the soils of China see: V. Kovda, *Soils and Natural Environment of China*, US Jnt. Publ. Res. Service, Oct. 1960. See also: I. P. Gerassimov, Sov. SS, 1, 1958, 3–12

15 A. A. Rode, *Soil Science*, IPST, pp. 468–72

16 R. D. Krebs and J. C. F. Tedrow, SS, 85, 1, 1958, 28–37, and R. W. Simonson, PSSSA, 14, 1949, 316–9

17 R. Brewer, CSIRO, Soil Pub. 5, 1955

18 C. B. England and H. F. Perkins, SS, 88, 5, 1959, 294–302

19 Y. N. Afanasiev, *Basic Outlines of the Earth's Soil Cover*. Bieloruss. Akad. Nauk., Minsk, 1931

20 V. M. Fridland, Poch., 7, 1962, 77–81

21 S. V. Zonn and Li Chin-Kvey, Sov. SS, 3, 1961, 245–52. Also Y. Hseung and M. L. Jackson, PSSSA, 16, 1952, 294—7

22 K. D. Nicholls, Reconnaissance Soil Maps of Tasmania, Sheets 46–7, Quamby-Longford, CSIRO, 1959. On New Zealand soils see *Soil Survey Method* 1962, p. 155–7 and *A Descriptive Atlas of New Zealand*, Wellington, 1960, pp. 28–37 and maps 12–13

17

Soils of intertropical areas

THERE is great scope for soil variation in intertropical areas, even if arid and mountainous areas are excluded. However, certain unifying influences exist—high temperatures (>25°CØ), high humidities and heavy rainfall. Equatorial regions have diurnal rain; tropical areas, dry seasons of varying duration—1 to 4 months. Intensities vary, but amounts are high (>1 mØ), reaching 5 m in mountains. Large areas have long, intense dry seasons (>4 months), with lower rainfalls (0·5–0·8 mØ), as in the Rhodesias and much of Kenya, and in the Sudan–Sahel zone at 12–18° N.

The usual contrast of dry and wet sites obtains, forming a catena from rocky crests to lowland swamps. There is, too, great variety of parent materials—the hard rocks of shields and large areas of recent volcanic materials. Plant growth is rapid, pausing only in the dry season. Organic matter has a rapid turnover, for at >25°C the rate of decay of litter tends to be greater than the supply,[1] while termite and worm activity are intense.

In intertropical areas alteration of rocks is usually deep seated and intense, producing clay-rich weathering mantles (Fig. 5) up to 100 m deep. Thus the distinction between soil, weathering mantle and parent material is hard to draw.[2] The agronomist would regard the soils as a thin superficial layer, impoverished, and only the climate responsible for luxuriant plant growth, unless deep rooted species occur.

In tropical weathering, felspathic minerals are decomposed intensively but slowly by physico-chemical means; with podzolization they are quickly dissolved by biochemical means. Thus simple clay minerals (kaolinite) form from the relatively resistant materials remaining after weathering and the removal of bases and alkalies. Only free Fe and Al oxides, and oxides of Ti, Cr and Ni are left, along with fine quartz and kaolinite, to form soil[3] (see Fig. 3). This process has been widely known as *laterization*; more lately as *ferrallitization*;[4] and the fully developed tropical soil as a *ferrallitic* (Ft) soil. The Fe oxides give the soils their reddish colour, enhanced on ferromagnesian rocks or on well-drained sites.[5]

Rock type is an important factor in tropical soil differentiation.[3] On neutral or alkaline (ferromagnesian) rocks, silicates are completely hydrolyzed,[6] with the 'individualization' of silica and Fe and Al oxides or hydrates, to form limonite ($2Fe_2O_3.3H_2O$) and gibbsite ($Al_2O_3.3H_2O$) (Fig. 3). In more acid milieu silicic colloids disperse, positive sesquioxides remain and kaolinitic rather than ferrallitic soils form.[7,8] On well drained sites on non-acid rocks, therefore, Si is removed along with bases into the drainage waters, and Ft soils, rich in free Al, are formed. Metallic oxides of Ti and Cr are stable and accumulate near the surface, Fe moves relatively easily and accumulates at the soil water table as *concretions*, thin *lenses*, or *soft compact masses* capable of hardening on dehydration.[4] Densest iron masses form on fine-grained basic igneous rocks, as *pisolitic laterite*,[9] and crusts are invariably deepest on Fe-rich rocks.

Kaolin-rich soils (*kaolisols*)[10] occur on wet or poorly drained sites if the Si is not completely removed by leaching on acid igneous rocks. These soils have recombined Si and Al, are less rich in free Al and some have Si accumulation at depth. Desilification of these materials forms *bauxite*, as on old coastlines in N. Surinam—or in the Nassau Mountains, with a crust having 63% Al_2O_3.[11]

In areas with a more prolonged dry period, rock weathering is less intense. Only Fe oxides are liberated and there is no free Al. The Al and Si remain combined in clay form and so *ferrisols* (Fc) or fersiallitic (Fg) soils occur.[7]

Tropical black soils. If drainage in humid tropical areas is impeded then *gleys* and *groundwater laterite* occur in areas of acid rocks, but, near to basic rocks or newly-weathered rocks, leached colloidal Si and bases accumulate in depressions forming *black tropical soils* (Fig. 13), dominated by *montmorillonite*.[12] Black calc-gleys are also known, of slightly different composition.

Black tropical (margalitic) soils are tropical pedocals, dependent on site and on base-rich water or parent materials for their formation. They are not chernozemic, for they have much less humus, different structure and other contrasts with the more northerly soils.[13] Their black colour derives from humus-montmorillonite clay complexes, humus contents being only $0.3\%–3.5\%$.[14] Plant debris layers are rare; vegetation is tall grass, low tree savanna, with $0.4–0.9$ mØ rain and a 4 months' dry season. Thin granular A2 and lighter calcic A3 may be discerned. In the compact prismatic B horizons, clay contents reach $50\%–90\%$, silt+clay contents are $>85\%$ (Fig. 5, 7), causing swelling in the wet season and drying and cracking in the dry season.[15] The soils are self mulching in the manner of *gilgai* and have very indistinct horizonation. They are saturated with Ca and Mg, and have a pH >8. Calcic and gypsic concretions may be present, and lime-rich margalitic soils may be grey rather than black.[15] Though fertile, the productivity of these soils is limited by their instability and by a tendency to flooding in the wet season (Figs. 5, 7 and 13, 6).

Black tropical soils have many regional variants—the *regur* on Deccan basalts, covering 200,000 sq miles, where grey hydromorphic *dhankar*, black clayey *karail*, and y/b lateritics form a catena.[13] Other forms are *tirs*;[16] the *massape* of Sao Paulo, the *margalitics* of the drier lowlands of E. Indonesia, the *black earths* of Australia,[17] the *black turf* of Transvaal,[12] the *makande* of Nyasaland, the calcic and non-calcic *black cotton soils* of Tanganyika, the *gravinigria* of Angola, lastly the *smonitzas* of central Yugoslavia and the *black soils* of Andalusia. In the tropics they are associated with ferritic red loams; in cooler latitudes with red-yellow podzolics or terra rossa. Their SiO_2/R_2O_3 ratio is c 4; that of tropical red soils is <2, while the C/N ratio of their surface horizon is lower (c8).

Tropical red soils. Rougerie,[7] reviewing interpretations of intertropical pedogenesis, wryly remarks that 'laterite' and 'a red colour' have become the sole symbols of our ideas on tropical soils. He distinguished four weathering 'milieu' of decreasing intensity:

(1) soils dominated by free Al and Fe—*ferrallitization*,
(2) soils dominated by Fe alone—*ferritic* soils,
(3) soils dominated by bases—black margalitic soils,
(4) soils formed by recent physical weathering.

These four are all capable of subdivision, as we have already seen in the case of margalitic soils. There are, too, various organic soils and eroded variants of many mineral soils.

The first two milieux have caused most controversy, and were once termed 'lateritic' or 'latosols'. The term 'latosols' may just be worthy of survival as an *omnium gatherum*; 'lateritic' could be restricted to soils with a hardened crust, or an indurated horizon or with a well-developed concretionary horizon.

The laterite horizon. The term *laterite* was suggested by Buchanan[18] in 1807, '. . . *a soft, earthy mass, full of cavities, contains a very large quantity of iron in the form of red and yellow ochres and hardening in the air and can be used for building*'. Though this statement confounds those recent writers who would limit 'laterite' to an already hardened, perhaps fossil, layer, this must be done to avoid confusion.[19]

The laterite crust is not necessarily the end product of soil evolution as implied in many studies. In general a material must develop with a high free iron oxide content.[9] For such enrichment iron can come from many sources, and the hardening of a soft saprolitic mass to a crust can be achieved through a number of varied causes. In addition to the concept of an end product, one of the causes is local and *anthropogenic*—the result of forest clearing and rapid hardening; one *paleogeographic*—drying by climatic change; and a third—the sequential or *geomorphic* cause—achieved by valley downcutting, sited high and dry on several high level geomorphic surfaces, and thickest on the oldest surface. The oldest laterites, therefore, are often quite unrelated to present climates.

D'Hoore[20] has distinguished between *relative* and *absolute* Fe accumulation in Ft soils, both capable of producing crusts. The *relative accumulation* of iron is a zonal chemical phenomenon resulting from *ferritization* on peneplains forming a saprolyte which hardens on exposure by bush clearing, forest firing, or sheet erosion of the overlying protective, shallow, bleached, but 'dusky' A layers. Crusts have been known to form in 30 to 60 years in West Africa from such causes, longer after rejuvenation of streams.

The second cause, an *absolute increase* of iron hydroxides, is the result of geomorphic or site factors, where oblique or lateral seepage from upper slopes adds Fe to soils on lower sites.[21] When these are oxidized, footslope crusts form. These are mainly located in drier areas or the wooded savanna, and, along with fossil crusts, form the majority of intertropical laterites.[7]

The iron-enriched saprolyte is usually nearer the surface the farther away from the equator;[9,22] and north of 14°N only fossil crusts are found. Peneplain, pediplain and capillary laterites are all found in the Ilorin plateau of southern Nigeria.[23]

Pendleton[24] has defined a laterite layer as '*a vesicular, concretionary, cellular, vermicular, slag-like, pisolitic or concrete-like mass, consisting chiefly of ferric oxides with or without quartz and minor quantities of Al and Mn.*' Variations of colour and composition occur. Crusts are dark red or violet if iron-rich; light red (rosé) or pink if bauxitic; greenish if rich in Al and phosphates; and grey or 'beige' if silicates are still present in the parent saprolyte.[25]

Perhaps it would be best to abandon the term 'laterite' for a native term such as 'murram'. Other native terms are 'khoai' (W. Bengal)[26] and 'bowal' (Guinea).[27] *Plinthite* is a modern term.[10]

Laterite layers are therefore distinct physical layers in *some* tropical soils. Attention must now be paid to the zonal ferrallitic and ferritic soils (from which crusts may form) studying their morphology; qualifying the idea of deep, clay-rich, red tropical soils with the thought that they can be shallow, yellow or red on acid rocks, brown on basic rocks, as well as grey or black. Variations in morphology, texture, and profile may be explained by reference to parent material, climate, site, vegetation and degree of development as shown in the well known diagram by Mohr.[3] A simple textural ratio[28]—silt/clay—is also a possible index of degree of weathering of tropical soils.

The tropical soil profile. Aubert[4] provides a profile of a deep Ft soil (Fig. 13, 1 and Table 12), which, in its upper part, is not *in situ*, but a colluvial deposit, interstratified with heavy-mineral layers and stone lines derived by sheet-wash or macro-organic sorting. This upper part is bleached or podzolized. Subsurface iron-enriched BL horizons, though developing *in situ*, are affected by soil solutions derived from upslope or by lateral seepage, which is intense if the lower BL3 are impermeable at 1–2 m depth. The subsoil—the speckled clay—'Bm' at 6 m and the CR horizons (9–14 m) are *in situ* and quartz-rich. The soil water at these depths in no way resembles the atmospheric water.

TABLE 12

A FERRALLITIC PROFILE, DAKPADOU, IVORY COAST (5°N 6°W)

Ferromagnesian gneiss on gently undulating terrain. 1·7 mØ rain; 27°CØ (23°–32°C); ombrophilous subequatorial forest (Fig. 13, 1):

cm

		cm	
(i)	LF	25–0	Bed of leaves, twigs and branches, pH 5.
(ii)	A(h)	0–35	Slightly humified greyish-brown gravelly fine sand.
	A(s)	35–50	G/b gravelly clay, hard, round Fe concretions and stone line.
	A3	50–80	Reddish g/b gravelly clay.
			Layer ii has pH 4·6, 20% SiO₂, 25% Al- and 33% Fe-oxides.
(iii)	BL1	80–110	Compact brick-red clay, softer concretions.
	BL2	110–175	Very compact, with bands of red or r/b crusts, anastomosing to enclose ochrous earthy masses.
	BL3	175–650	Compact non-hard red clay (50% kaolinite). Beige or grey 'eyes'; quartz and grey 'pipes' at depth. Some hard concretionary material at 2 m.
			Layer iii has pH 4·9; 22%–31% SiO₂, 24%–32% Al, 19%–35% Fe.
(iv)	B(m)	650–840	Speckled clay, more grey 'pockets' and powdery felspars. pH 5·1; 35% Si; 25%–30% Al, 14% Fe.
(v)	Bsi	840–900	Ochrous-brown, quartz-rich, with core stones. Quartz crushes to a fine powder.
	CR1	900–1200	Gneissic sand (arène). Quartz and friable white felspathic material in original shape.
	CR2	1200–1400	Slightly weathered rock, pH 5·2; with 37% Si, 33% Al, 11% Fe.

Note: The suffix BL has been proposed for a non-hardened ferric horizon;[3] Lcr or Lr for a surface crust. The suggestions for horizon nomenclature are by the author, and not from the original.[4]

The five divisions of the Ft profile may now be discussed with reference to their variations with climate, parent material and site (see also Fig. 5, 1–3 and Fig. 13).

Organic layers (i) are normally thin, and usually thicker (c 0·4 m) on fine textured soils than on sands (0·1 m). Fallen branches and trunks quickly decompose, and litter at a rate of 1·3% per day.[29] Ah layers also have low humus contents (1%–1·5%). Humus is moderately rich in N, with C/N ratios of 10–16, decreasing to c 3 in the mineral soil.[30] Clearing and cultivation rapidly reduce the amount and C/N ratio of humus.[29]

Intense decomposition gives various organic materials. Soluble colourless humus is formed by termites, aiding aggregation;[31] micro-organisms produce darker, less-soluble inert humus, neutralized by plant ash. Humic acids combine with Ca and are destroyed at 28°C,[32] while fulvic acids achieve podzolization of the A layers and accumulate, increasing with depth.

Humic ferrallitic soils form in perhumid areas, in high altitude grasslands, or under moss forest, with *c* 1 m deep organic layers, especially on volcanic or basic rocks. Contents of 10% humus have been noted in the W. Cameroons (3·4 mØ rain) and >20% in Hawaii, with *c* 8 mØ rain,[33] where upper mineral layers are purplish silt loams, rich in heavy minerals, underlain by y/b bauxitic clay of normal density. Similar humic Ft soils are found in Colombia, Honduras, Liberia and N.E. Madagascar. Their C/N ratio is *c* 16–30.[34] (See also Fig. 6.) Such soils occur above 1,500 m.

The upper mineral horizons (ii) are greatly leached and greyish, if not eroded, and have some Fe concretions. Under equatorial conditions, or on ill-drained plateau and peneplains on acid rocks, the upper layers are very light, forming *pallid zones* or *grey latosols*.[35] The grey horizons result from reduction or removal of hydroxides, or by biogenic accumulation of silica as phytolites.[34] On many peneplains or in depressions, variation of a water table from 0·5 m depth in the wet season to 2 m in the dry season gives capillary Fe enrichment and formation of laterite at shallow depth (Fig. 13, 7)— the *ordinary groundwater laterite*.[5]

The red colour of tropical soils need not indicate a high iron content nor grey the lack of it; for Fe in hydrated form in yellow or wet grey soils may exceed the iron content of red soils.[36,37] The degree of hydration is the cause of the colour change.

Layer iii attains 10 m thickness. The BL 1–2 horizons are most compact and rich in resistant metallic hydroxides. Usually red, they can be yellow or ochrous. The lower part—BL3—is most kaolin-rich, with traces of illite and clay-sized quartz. Larger quartz crystals collapse to fine powder on pressure.[4] The BL3 has a stable 'nutty' structure, with some soft Fe concretions. Its upper parts harden progressively downwards only to *c* 2 m depth.

Layer iv is a moist, compact, mottled (speckled) clay. If it is ever caused to dry out, scoriaceous (cellular) laterite forms. It is best developed in very moist coastal lowlands on acid rocks, and the

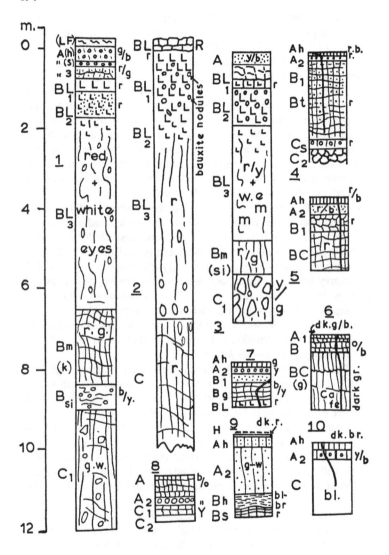

mottled clay is absent on drier lowlands and poorly developed on basic rocks.

Layer v—CR—is thickest on acid rocks (<5 m). It is the true parent material, or 'zone de départ' in the cautious French terminology,[4] and has large beige pockets and is porous. The original rock structure is visible, with concentric weathering skins, and pH is higher than in other layers.

The deep reaching and intense alteration of rock minerals in tropical soil profiles has been explained by acid hydrolysis—the replacement of cations by H^+-ions,[38,39] or of the O^- ions in the crystal lattice by the OH^- radical.[7] Bases and soluble Si are removed while the metallic hydroxides concentrate at appropriate depths. Free Fe oxides form hematite or turgite as nodules or crusts, Al oxides form hydrated bauxitic (gibbsite or boehmite) masses.

Erhart[40] took this a stage further in his 'Theory of Biorhexistase' for under forest 'rhexistasie' there is complete ferrallitization; while on climatic change to drier savanna, hydroxides are dehydrated and form crusts. In deserts 'biostasique' conditions obtain, with alkaline, usually regosolic or litholic soils.

Aubert's approach[41] is of deep Ft soils under forest; shallower ferruginous soils under savanna, which are less acid and without free Al. They can be leached—ABtC-soils, or non-leached—A(B)C-soils. Both can form crusts. The illuviated clay in the Bt, at c 1 m depth, promotes hydration and yellowish A and B1 horizons result. Eventually the silica in the Bt is dissolved and ferrallitization commences.

Tropical pedogenesis may also be viewed as a sequence of soils, with differing proportions of clay and resistant minerals dominating

---

Fig. 13   Some intertropical soil profiles (width of column=50% clay):
1 Ferrallisol, Dakpadou, Ivory Coast
2 Angadipuram, lateritic ferrallisol
3 Deep lateritic ferrisol, Ivory Coast
4 Fersiallitic with textural B, Angola
5 Red tropical loam, Uganda
6 Tropical black soil, Kenya
7 Ground water laterite, Ghana
8 Tropical rendzina, West Irian
9 Tropical podzol, Cameroons
10 Andosol, São Tomé

at successive stages. At present, now that kaolinite has been shown to alter under humid tropical conditions, many views converge on a sequence of multiple hydrolysis,[7] proposed in 1926 by Harassowitz,[39] illustrated here by breakdown of orthoclase:

$$K_2Al_2O_36SiO_2[+H_2O] \rightarrow H_2OAl_2O_36SiO_2 + 2KOH \text{ (leached)}$$
$$[+H_2O] \rightarrow 2H_2OAl_2O_32SiO_2 \text{ (ie kaolinite)} + 4 SiO_2 \text{ (removed)}$$
$$[+H_2O] \rightarrow Al_2O_3 + 2Si(OH)_4.$$

Within an Ft profile there are internal 'microclimates' favouring each step in this multiple hydrolysis and, if not in one profile, then in soils on adjacent but slightly different sites.[7]

The soil moisture regime in tropical soils is not a rapidly moving leaching regime, as in temperate soils, but that of a much altered, concentrated, slow moving liquid. It achieves intense hydrolysis in the clay-rich mass and is derived from three main sources: rainfall; added seepage if on a suitable site; and vapour transfer. Diurnal vapour absorption may be 3 mm to depths of 1 m—90% of total movement in the dry season, 2%-6% in rainy periods.[42]

Intertropical soils and rainfall

1. Areas with >2 mØ rain, with brief dry seasons, have leached Ft soils on well drained sites. At an early stage of development, the base content of parent rocks has an effect. With time the mantle deepens, an acid front moves to depth and plants cease to extract bases. Little ash is returned to the surface and Si and Al combine in the acidified layer. Superficial podzolization occurs,[43] bauxite may form from kaolinite and then gibbsite $Al_2(OH)_6$.[7,44] In ill-drained areas kaolinite-rich gleys (hydrokaolisols),[45] humic latosols, or groundwater laterites are formed.

2. Areas with c 1·5 m and a short dry season have crusted Ft soils (Fig. 13, 2). Hydrolysis continues in the dry season at depth, for subsoils are always moist. Dehydration of the BL1 occurs in the dry season, while annual or long term variations in the depth of the drying front account for the anastomosing red layers in the BL2. Sys has termed such soils *hygro-xerokaolisols*.[45]

3. Subhumid tropical climates with c 1 mØ rain and a longer dry period have shallower *ferrisols*.[46] Al and Si are stable but some Fe is leached in wet periods. In the dry season sesquioxides accumulate. With time the soil is progressively dried and a thin hard crust forms.[7]

4. '*Xeroferric soils*' occur with <0.8 m$\varnothing$ rain and a long dry season. Bases and some Fe are separated out, and much residual clay forms as *ferrisols* if rainfall is c 0.8 m. With 0.5 m$\varnothing$ rain friable, permeable r/b *ferruginous soils* (Fg) form (Fig. 13, 5).[45] With 0.3 m$\varnothing$ rain alkaline, solonetsic brown soils form, transitional to the chestnut and reddish brown semi-desert soils, all on aeolian deposits.

The relief factor is the second major key to an understanding of tropical soils (Fig. 14), influencing soil moisture regime, and determining crust formation or pallid layer development on dissected peneplains. The factor is integrated in the catena, and since Milne's original work many catenas have been reported.[47,48,49] (See Fig. 14, 5).

Tropical soils may also be related to dated geomorphic surfaces. Upland S.W. Kenya[50] shows an early Cretaceous (post-Gondwana) peneplain with <2.5 m$\varnothing$ rain, mature Ft soils on well drained sites, and pallid soils in poorly drained areas. The Miocene African peneplain has less evolved Ft soils and ferrisols, with thinly crusted ferritic loams in its drier parts. Low-lying undulating Pleistocene surfaces with a subhumid climate have well drained y/r ferrisols or grumusols. Intense Pleistocene erosion of the uplands produced extensive lithosol areas, especially on old sedimentary rocks. Valley downcutting left the margins of older peneplains high and dry, with deep, flanking plinthite, crusts.

In the Congo,[51] 'recent tropical soils' are related to active slopes around eroding inselbergs and sugar loaf features. Deeply altered soils on the oldest pediplains and on undissected laterite-covered plateau are termed *ferralsols*, with gibbsitic B horizons and no clay skins. Younger pediplains have regosols if undergoing active transport; their lower parts, with clay-rich debris, have *ferrisols*, with clay skins in their Bt horizons.

Tropical soils—reaction and base content. A classification developed in Ghana[52] has gleys and peats as hydromorphic; margalitic and podzolized soils as intrazonal soils. There are three, perhaps four, zonal divisions: *basisols* on ferromagnesian rocks, with pH c 7 and high CEC; *ochrosols* with intermediate base saturation $(25\%-75\%)$, pH 6–7, under semi-deciduous forest with c 1 m$\varnothing$ rain; *intergrades* are very similar and, with ochrosols, are termed 'cocoa soils'; while *oxysols*, with pH <5, occur with >2 m rain and no dry season. Such are yellowish, allitic, and best developed on quartzites in S.W. Ghana. Their base saturation is <15% and they are not very productive, being used for rubber and oil palm production. Similar soils occur in N. Borneo[53] and in the Amazon lowlands.[54]

In ochrosols, rubefication is most marked on dry sites, with red loams; while moist sites have y/r clay loams. A distinction is also made between 'plateau soils' with many concretions and crusts, and 'non-concretionary soils' on piedmont areas.[52] Groundwater latosols (Fig. 13, 7) and gleys are their hydromorphic associates, the former the shallowest and least productive of all tropical soils. In N. Ghana the valleys have savanna gleys, with higher base status. On basic rocks in the savanna (7°N) *rubrisols* and *brunosols* occur, with alkaline surfaces. They are ferruginous soils, with reserves of primary minerals. Margalitics are common at 9°N, also in the drier Accra-Keta plains in the south-east.

In the Ivory Coast,[55] a latitudinal sequence has been presented:

1. The South, with >1·8 mØ rain, gallery or rain-forest and red-yellow Ft soils (>15 m deep), not dehydrated and crustless if the forest cover is intact. Degradation is by acidification rather than crusting, with very strongly leached soils on perhumid plainlands.

2a–b. The 'Celtis' Plains and Plateau, with c 1·5 m rain and semi-deciduous forest, have less deep (10 m) Ft soils with crusts on dry sites. Entrophic brown tropical soils occur on ferromagnesian rocks.

2c–3. A transition at the forest-savanna boundary (6°N) has forest on clayey Ft soils and savanna on leached or crusted red sandy soils. Fc soils dominate the Agneby and Bandama lowlands.

3. Between 6° and 10½°N, the savanna woodland, has <1·3 mØ rain and 6 months' dry season, with many Fe crusts under savanna and 4 m deep ferrisols under the remaining woodland. Such soils are also widespread in S. Senegal and resemble the red-brown soils of Ghana. Immense lateritic crusts also surround the massif of Fouta Djalon (11°N, 12°W) in Guinea at heights >800 m.[56]

4. The Sudan zone, north of 12°N, in Mali-Volta, with 8 dry months and 0·9 m rain from June to September, has Fg soils, fossil crusts, ground water laterites and vertisols in the Souron basin.

5. The Sahelian zone (0·2–0·5 m rain), has fossil crusts but no zonal laterite. In the south the Fe crusts are gullied, in the north covered by wind-blown sand. Saline and r/b subarid soils prevail.

Asiatic intertropical soils. True laterite soils cover only 5000 sq miles in southern India.[8] Thin and gravelly at high levels, they elsewhere range from heavy loams to clays. *Red krasozemic loams* cover far greater areas—300,000 sq miles. In Burma,[57] laterites and Ft soils are also rare, confined to S. Arakan and Tenasserim, and laterite is rare above 150 m altitude. The zonal soil of Burma is also a red loam in the evergreen forest and a y/b or brown soil in the monsoon areas. The soils of S.E. Asia are classified by Dudal.[57]

In Puerto Rico,[58] latosols have been characterized by their geomorphic occurrence, while in S. America,[54,59] dark latosols occur in humid areas or on base-rich rocks; red latosols on acid materials and brown latosols on ash or basaltic terrain. Yellow forms are common in ill-drained areas, while dark clayey *talpatete* are margalitic. Regional contrasts in S. American soils have been described by Bennema et al.[59]

Tropical soils have also been studied in Angola, where Bothelo[25] distinguishes between Ft soils and 'ferrisiallitic soils', which are less weathered and on the forest-savanna boundary. They are rich in kaolin and micaceous clays, with high CEC and original mineral contents. They correspond to ochrosols.

Intertropical soils are complex in profile and origin, and variable in colour, depth and distribution. There is ample scope for at least six major sub-zones and also hydromorphic, regosolic and paleogeographic forms. Many complex geomorphic influences act or have acted, so tropical soil surveys use the concepts of catena and association in mapping soils, and pay more attention to geochemical rather than morphologic criteria in describing profiles.[60]

The work of correlation and synthesis of tropical soils has hardly begun. D'Hoore[46], in a convincing and widely accepted scheme, has proposed a five-fold division of the major intertropical soil forms:

1. *Ferrallitic soils.* Intensely weathered deep soils with no original rock debris, little or no silt (silt/clay ratio <0.15), low base exchange capacity, much immobile clay and little or no plasticity. Kaolinite dominates, with some gibbsite (if gibbsite is dominant then *alumisols* occur). *Humic ferrallitic soils* form a subgroup. Ft soils are usually red, but can be yellow on clay-rich parent materials. Ft soils are divided into *strongly ferrallitic* with an SiO_2/Al_2O_3 ratio <1.3; *normal* $(1.3–1.7)$; and shallow (3.6 m) *weakly ferrallitic* $(1.7–2.0)$. They

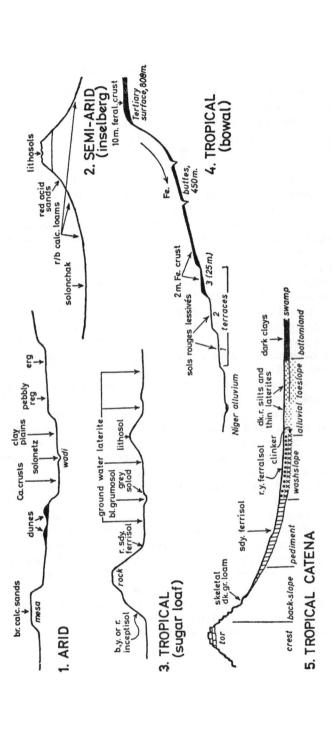

1. ARID

mesa — br. calc. sands — dunes — Ca. crusts — solonetz — clay plains — pebbly reg — erg
wadi

2. SEMI-ARID (inselberg)

lithosols — red acid sands — r/b calc. loams — solonchak
Tertiary surface, 808 m.
10 m. feral. crust

3. TROPICAL (sugar loaf)

b., y. or r. inceptisol — rock — r. sdy. ferrisol — ground water laterite — bl. grumosol — grey solod — lithosol

4. TROPICAL (bowal)

Fe. — buttes, 450 m. — 2 m. Fe. crust — 3 (25 m.) — 2 — 1 terraces
sols rouges lessivés
Niger alluvium

5. TROPICAL CATENA

tor — skeletal dk. gr. loam — sdy. ferrisol — r.y. ferralsol — clinker — dk.r. silts and thin laterites — dark clays — swamp
crest | back-slope | pediment | washslope | alluvial toeslope | bottomland

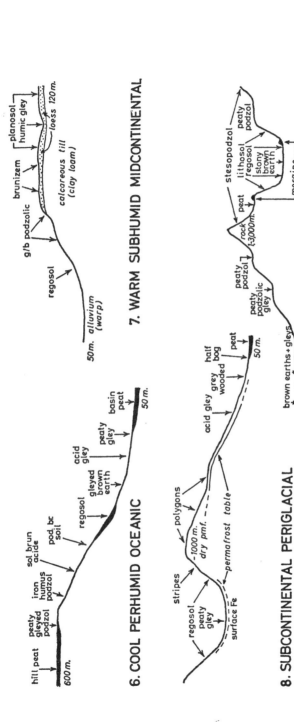

Figs. 14   Soil associations of representative climo-geomorphic units, in tropical and in temperate and arctic areas. Horizontal scales may be deduced. Indications of vertical scale are given where possible

are porous, despite their massive structure and high clay content. They have A, BL, Bm, CR profiles, while *lateritic soils* have Lcr, BL, CR profiles. (See Fig. 13, 1 and 2.)

2. *Ferrisols* are less weathered, shallower and younger soils, having more silt (silt/clay ratio $0.5–0.15$), clay coatings in a structural B, and kaolinite, illite and traces of montmorillonite. The Si/Al ratio is *c* 2. They may have lessivé or non-lessivé forms, and ABC, ABtC, or A(B)C profiles. Thin crusts may be present on sands or in deforested areas.

3. *Fersiallitic soils* (Fig. 13, 4) are rich in free Fe, gibbsite is absent and Si/Al ratios exceed 2. They have AC, A(B)C or ABtC profiles, the B and C with higher pH and *c* 40% base saturation. The effect of parent material is quite marked in these soils; they are often silty, and are transitional to:

4. Young dark brown tropical soils (or r/b or y/b) are stony, with much primary material, textures varying with parent materials. Kaolinite is lacking and illite often dominates. They have AC or AB(t)C profiles with either iron nodules or Ca deposition at depth.[61]

5. Divisions comprising litho and regosolic, calcimorphic, vertisolic, halomorphic, hydromorphic and organic soils found in intertropical areas.

These words,[4] applicable to almost all soils, in all places, and at all times, may be quoted as a final comment on tropical soils.

'*Ces sols, souvent peu fertiles, sont utilisables dans la majorité des cas, et nous devons chercher à accroitre leur productivité. Ils sont très fragiles et en quelque années des méthodes agricoles défectueuses peuvent les rendre pratiquement stériles. Il est donc de première importance que leurs caractères, leur formation, leurs réactions soient mieux connus, sinon leur mise en valeur risque de conduire à des échecs. Nous n'avons pas le droit de laisser inutilisées ou mal utilisées de si vastes étendues capables de fournir d'immenses quantités d'aliments et de produits de première nécessité pour l'homme.*'

1 N. Ching Wen, Poch., 5, 1961, 34–42

2 L. Berry and P. Ruxton, JSS, 10, 1959, 54–63

3 E. C. J. Mohr and F. A. van Baren, *Tropical Soils*, 1954. See also H. Vine, CBS., Tech. Comm. 46, 1949, 22–9, and G. Pedro, CRAS, Paris, 247, 1958, 1358–60

4 G. Aubert, CRICSS, 5, Leopoldville, 1954, I, 103–18. Profiles from N. Vietnam are given in V. M. Fridland, Sov. SS, 12, 1961, 1323–37

5 N. Leneuf and G. Rion, *Sols Afr.*, VIII, 3, 1963, 451–62

6 D. Carroll and M. Woof, SS, 72, 2, 1951, 87–99

7 G. Rougerie, CRAS, Paris, 246, 1958, 447–9; also Inf. Géog., 23, 5, 1959, 199–205; and *La Faconnement actuel des Modelés en Côte d'Ivoire forestière*, Mém. IFAN., 58, Dakar, 1960

8 K. V. S. Satyanarayana and P. K. Thomas, J. Indian SS, 9, 2, 1961, 107–18 and 10, 3, 1962, 211–22

9 L. T. Alexander et al., USDA Tech. Bull. 1282, 1962; also Adv. Agron., 14, 1962, 1–60 and R. Maignien, *Sols Afr.*, IV, 4, 1959, 4–41

10 C. Sys, C.R. III Interafr. Soils Congr., Dalaba, 1959, 303–16

11 J. P. Bakker, Rév. Géom. Dyn., 1959, 67–84

12 C. R. v. der Merwe and H. Heystak, SS, 79, 2, 1955, 147–58

13 H. Shiva Rau and S. Kasinathan, JSS, 2, 1951, 61–6

14 V. S. Ramachandram et al., Proc. Ind. Acad. Sci., 50A, 1959, 314–22

15 R. W. Simonson, JSS, 5, 1954, 275–88; also R. Dudal, SS, 95, 4, 1963, 264–70

16 H. Oakes and J. Thorp, PSSSA, 15, 1951, 347–54

17 J. S. Hosking, Tr. Ryl. Soc. S. Aust., 59, 1935, 168–200

18 F. Buchanan-Hamilton, *A Journey from Madras through the Countries of Mysore, Canara and Malabar*, London, 1807, vol. 2, p. 441

19 C. G. Stephens, JSS, 12, 1961, 214–7

20 J. L. D'Hoore, Med. Landb. Hogesch. Gent, 19, 1954, 98–248. See also R. L. Pendleton, AAAG, 37, 1947, 50–1 (Peneplain and slope laterites of Siam)

21 R. Maignien, B. Ass. Fr. Ét. du Sol, 5, 1960, 244–64 and TICSS, 7, 1960, V. 24, 171–6

22 P. Gourou, *Tropical World*, Longmans, 3rd ed., 1961, 21–2

23 J. W. du Prees, Bull. Agr. Congo Belg., 40, 1949, 53–66

24 J. A. Prescott and R. D. Pendleton, *Laterite and Lateritic Soils*, CBS., Tech. Comm. 47, 1952

25 Mem. da Junta Invest. do Ultramar, 9, 1959, *Carta Geral dos Solos de Angola*; (1) Distrito de Huila. Also (2) Huambo, 27, 1961, Lisbon

26 There are many other local names for 'laterite'.

27 A. Aubreville, Agron. Trop., 2, 1947, 339–57. The term 'bovalisation' has been applied to the isolation of laterite-capped mesa-forms by pedimentation

28 A. R. van Wambeke, JSS, 13, 1, 1962, 124–32

29 P. H. Nye, Plant and Soil, 13, 1961, 333–46

30 H. F. Birch and M. T. Friend, JSS, 7, 1956, 156–67

31 A. de Craene, TICSS, 6, 2, 1956, 707–12; and D. W. Duthie, E. Afr. Res. Inst., Amani Rept., 5, 1947

32 G. Bachelier, Pedobiologia, 2, 1963, 153–63. With regard to temperatures, forested land (surface c 28°C) is usually cooler than the atmosphere (36°C) while bare soil surfaces approach 50°–54°C. (P. Vageler, *Tropical Soils*, Macmillan, 1933) also R. Maignien, *Sols Afr.*, VI, 1961, 214

33 T. Tamura et al., PSSSA, 17, 1953, 343–6

34 R. Chaminade, Bull. Agr. Congo Belg., 40, 1949, 303–8

35 J. P. Watson, Nature, 196, 1962, 1123–4 and S and F, 1, 1962, 1

36 K. Utescher et al., ZPDB, 40, 1948, 206–37

37 J. M. Oades, S and F, 26, 2, 1963, 69–80

38 J. B. Harrison, *The Katamorphism of Igneous Rocks under Humid Tropical Conditions*, Imp. Bur. Soil Sci., 1933–4

39 H. Harrasowitz, in *Handbuch der Bodenlehre*, III, Berlin, 1930, 387–436

40 H. Erhart, *Traité de Pedologie*, II, Strasbourg, 1938, and *La Genèse des Sols en tant que Phenomène Géologique*, Masson, Paris, 1956

41 G. Aubert, in *Rapports entre Sol et la Végétation*, Masson, 1960, 11–22 (ed. G. Viennot-Bourgin; Prem. Colloq. Soc. Bot. Fr., 1959)

42 V. Pagel, A. Thaer Arkiv, 4, 1960, 325–45

43 Hence, by some, podzolized ferrallitic soils are held to form in the presence of plants which are unable to reach the bases present in the subsoil; while lateritic soils form when plants are absent, or are removed

44 J. P. Bakker, Zeit. Geom. Supp., 1, 1960, 69–92

45 C. Sys, Pédologie, X, 1960, 48–116

46 J. L. D'Hoore, TICSS, 7, 1960, V. 2, 11–19

47 S. A. Radzwanski and C. D. Ollier, JSS, 10, 2, 1959, 149–68

48 W. E. Calton, TICSS, 5, 1954, 4, 58–61

49 A. H. Bunting, The Catena, S and F, 16, 1953, 331–4

50 J. W. Pallister, GJ., 122, 1956, 80–7; J. Thorp and E. Bellis, TICSS, 7, 1960, V. 46, 329–34; *Soil Map of East Africa*, 1:4m., Dir. Overseas Surveys, 299G, 1961

51 P. Jongen, TICSS, 7, 1960, V. 47, 335–46

52 H. B. Obeng, TICSS, 7, 1960, V. 35, 251–5; and P. H. Nye, JSS, 5, 1954, 7–21 and 6, 1955, 73–83. (The terms are after C. F. Charter)

53 P. H. T. Beckett and D. H. Hopkinson, JSS, 12, 1, 1961, 40–51

54 C. F. Marbutt and C. B. Manifold, Geog. Rev., 16, 1926, 226–76

55 N. Leneuf, *L'Altération des Granites Calco-alcalins et des Granodiorites en Côte d'Ivoire forestière*, Lang-Grandemange, Paris, 1959, pp. 210

56 A. Chevalier, Bull. Agr. Cong. Belg., 40, 1949, 1057–92 and ref. 9

57 V. G. Rozanov et al., Sov. SS, 12, 1961, 1338–45. R. Dudal et al., J. Trop. Geog., 18, 1964, 54–80

58 J. A. Bonet, TICSS, 4, Amsterdam, 1950, 7, 281–5

59 J. Bennema et al., TICSS, IV & V, N.Z., 1962, 493–506

60 R. Webster, S and F, 23, 2, 1960, 77–9

61 G. D. Anderson, *Sols. Afr.*, VIII, 3, 1963, 339-47. Students may be referred to the Maps and Explanatory Monograph on the *Soil Map of Africa: Scale 1 to 5,000,000* by J. L. D'Hoore, CTCA Pub. No 93, Lagos, 1964

Select bibliography

A. Older sources, not entirely superseded

E. Blanck, *Handbuch der Bodenlehre*, I–IV, Springer, Berlin, 1930
K. D. Glinka, *Treatise on Soil Science*, 1931, transl. A Gourevitch, IPST, Jerusalem, 1963
J. S. Joffe, *Pedology*, 2nd ed., New Brunswick, 1949
C. F. Marbut, Soils of the United States, *Atlas of American Agriculture*, USDA, 1935, Part III
E. Ramann, *The Evolution and Classification of Soils*, transl. C. L, Whittles, Cambridge, 1928
G. W. Robinson, *Soils, their Origin, Constitution and Classification*. 4th ed., Murby, 1949
A. A. J. de Sigmond, *The Principles of Soil Science*, Murby, 1938

B. Newer works concerned with soil geography or morphology

P. Duchaufour, *Précis de Pédologie*, Masson, Paris, 1960
R. Ganssen, *Bodengeographie*, Koehler, Stuttgart, 1957
I. P. Gerassimov, and M. A. Glazovskaya (*Fundamentals of Pedology and the Geography of Soils*), Gosud. Izdat. Geog. Lit., Moscow, 1960 (in Russian)
G. V. Jacks, *Soil*, Nelson, 1954
W. L. Kubiena, *Soils of Europe, Diagnosis and Systematics*, Murby, 1953
W. Laatsch, *Dynamik der Mitteleuropäischen Mineralböden*, Steinkopf, Dresden, 1957
E. Mückenhausen, *Entstehung, Eigenschaften und Systematik der Böden der Bundesrepublik Deutschland*, DLG Verlag, Frankfurt M., 1962
I. I. Plyusnin, *Reclamative Soil Science*, transl. I. Sokolov, Foreign Languages Publ. House, Moscow
A. A. Rode, *Soil Science*, Moscow, 1955, transl. A. Gourevitch, IPST, Jerusalem, 1962

C. G. Stephens, *The Soil Landscapes of Australia*, Pub. 18, CSIRO, Melbourne, 1961

S. A. Wilde, *Forest Soils*, Ronald Press, N.Y., 1958

Soil-Geographical Zoning of the U.S.S.R., IPST, 1963

Soil Classification. A Comprehensive System, 7th Approximation. Soil Survey Staff, USDA, 1960

B. T. Bunting, *An Annotated Bibliography of Memoirs and Papers on the Soils of the British Isles. Part I.* Geomorphological Abstracts, London, 1964

C. Texts on Agronomy

T. L. Lyon, H. O. Buckman, N. C. Brady, *Nature and Properties of Soil*, 6th ed., 1960

E. W. Russell, *Soil Conditions and Plant Growth*, 9th ed., Longmans, 1961

L. D. Baver, *Soil Physics*, 3rd ed., Wiley, NY, 1956

D. Works of reference

G. Plaisance, *Lexique Pedologique Trilingue*, Centre Documentation Univ. Paris, 1958

G. V. Jacks, R. Tavernier, et al., *Multilingual Vocabulary of Soil Science*, FAO, Rome, 1960

Index